Dialogues and Games of Logic

Volume 8

The Algebrization of Mathematics during the 17th and 18th Centuries

Dwarfs and Giants, Centres and Peripheries

Dialogues and Games of Logic Series Editors
Shahid Rahman   shahid.rahman@univ-lille3.fr
Nicolas Clerbout
Matthieu Fontaine

The Algebrization of Mathematics during the 17th and 18th Centuries

Dwarfs and Giants, Centres and Peripheries

Edited by

Davide Crippa

and

Maria Rosa Massa-Esteve

ISBN 978-1-84890-394-4

College Publications
Scientific Director: Dov Gabbay
Managing Director: Jane Spurr

www.collegepublications.co.uk

Original cover design by Laraine Welch

CONTENTS

ARTICLES

INTRODUCTION

DAVIDE CRIPPA

Institute of Philosophy, Czech Academy of Sciences, Czech Republic
`crippa@flu.cas.cz`

MARIA ROSA MASSA-ESTEVE

Departament de Matemàtiques. Universitat Politècnica de Catalunya-Barcelona Tech, Spain
`m.rosa.massa@upc.edu`

1   The algebrization of mathematics in 17– and 18–century Europe

Seventeenth century mathematics has been transformed through the interaction of three fundamental forces. The first one was the classical mathematical heritage of the sixteenth century, exemplified by the direct recovery in Greek and Latin translations of works by Euclid, Archimedes, Aristarchus and others; the second one was the "infinity revolution", that is, the extension of mathematics thanks to the use of infinite procedures and the study of geometric objects of infinite dimension; and the third one was the emergence of algebra and its use for solving problems.

The editors would like to acknowledge the support of the Czech Science Foundation (GAČR) Grant GJ19-03125Y, "Mathematics in the Czech Lands: from Jesuit Teaching to Bernard Bolzano" and the support of the project PID2020-113702RB-l00, "Mathematics, Engineering, Heritage: new challenges and practices (XVI-XIX centuries)" of the Spanish Ministerio de Ciencia e Innovación. The book was completed thanks to the European Union's Horizon 2020 research and innovation programme under the Marie Sklodowska-Curie grant agreement No 101024431-LEGITIMATH. The editors would also like to thank Alexander Reynolds and Jeff Palmer for their careful linguistic revisions, and Jane Spurr for her invaluable help with the editorial process.

The emergence of symbolic algebra as a formal language in mathematics is commonly referred in the literature as the "algebrization" of mathematics (see [Massa-Esteve, 2006]; [Massa-Esteve, 2020]). The process of algebrization involved a cognitive shift, namely a change in the way of thinking about mathematics, marked by the transition from explicit constructions and geometric explanations to the reliance on formulas and equations as the main way to represent mathematical entities and reason upon them (see [Massa-Esteve, 2001]). This process has been called by the historian H. Bos [2001, p. 10] "degeometrization of analysis". Consequently, algebrization also prompted the emergence and the establishment during the eighteenth century of analytical geometry and calculus (or infinitesimal analysis) as two autonomous mathematical disciplines.

Building upon Jacob Klein's seminal work ([Klein, 1976/1936]), the historian M. Mahoney (see [Mahoney, 1980]) argued that changes in mathematics pedagogy played a crucial role in facilitating evolution towards an algebraic mode of thinking in the early modern period. The diffusion of Ramus' didactic model, which prioritized analysis as a more fruitful and intuitive teaching method compared to the synthesis, as shown in Euclid's proofs, contributed to the adoption of algebraic formulas. Therefore, according to Mahoney [1980], the inclination toward algebra can be seen as an outcome of the imposition of this "modern" teaching method.

This pedagogical emphasis on analysis is also evident in textbooks from the eighteenth century. For example, according to Christian Wolff, the author of a successful mathematics course, it was imperative for those seeking a solid foundation in mathematics to study analysis. In Wolff's context, "analysis" encompassed both algebra and calculus, which made use of a symbolic notation. It served as a framework that encompassed generalized arithmetic and provided a method for discovering new mathematical results and proofs. Wolff stressed that analysis should be learned because it engenders concepts that surpass ordinary imagination and enables the condensation of lengthy chains of reasoning ([Wolff, 1717]; *Elementa Analyseos Mathematicae, Praefatio*). The pedagogical virtues of analysis, in its algebraic form, were recognized by other eighteenth-century authors as well. For instance, Leonard Euler translated the geometric derivation of a cannonball's flight into analytical language because it was "simpler, clearer, and of greater utility" ([Alder, 2010, p. 72]).

However, teaching and learning algebraic analysis posed epistemic challenges that are still relevant in contemporary teaching practices (see [Massa-Esteve, 2020]). Such challenges stem from the chief characteristics that Mahoney [1980] attributed to the algebraic mode of thought: the use of symbolic operations, focussing on mathematical relationships rather than objects, and freedom from ontological commitments. These characteristics account for the fact that algebraic thinking does not align with intuitive understanding. Algebra allows for the introduction and manipulation of entities that are challenging to define or work within the framework of classical geometry. For example, negative and imaginary quantities or objects in dimensions beyond three have posed difficulties as they lacked an intuitive representation. Similarly, infinitesimal analysis, as seen in the Leibnizian and Newtonian calculi, employs a symbolic system that enables the manipulation of impossible objects such as infinitesimals, which do not fit the framework of Euclidean mathematics. Given this lack of proper foundations, how could mathematicians and teachers of mathematics be sure that algebraic reasoning led to a correct, i.e. non contradictory and meaningful results?

The answer to this question is not obvious, and algebra and infinitesimal calculus had a share of critiques as soon as they were circulated. For Hobbes and for Barrow, for instance, algebra was merely a stenographic notation to abbreviate arithmetical or geometrical proofs, but did not constitute a form of scientific knowledge on the same level of arithmetic and geometry ([Mancosu, 1996, pp. 86–88]). Likewise, the reception of the Leibnizian calculus in countries like France and Italy had to overcome an initial reticence due to the mistrust toward the obscurities of the new symbolism, especially when compared with the geometrical methods of the Greeks (see, in particular [Mazzone and Roero, 1997] for the Italian case, [Mancosu, 1989] for the French one). We thus wonder how pedagogical advantages associated with the formalism of algebra and infinitesimal analysis, in terms of the economy of thought and expansion of our imagination, were eventually able to overcome the corresponding cognitive and epistemic difficulties.

3

In this connection, and despite the considerable research conducted on the algebrization of mathematics and its historical development, there are further avenues for exploring how mathematicians, practitioners, students, and other learners gradually accustomed themselves to the advanced mathematics that made used of symbolic notations. While not providing definitive answers, the contributions in this book hope to provide lines of inquiry for understanding the fundamental shift that brought the process of algebrization and its cognate disciplines of algebra, analytic geometry, and calculus (or infinitesimal analysis) to the core of modern mathematics.

One significant theme discussed throughout this book is the role of institutions in shaping the teaching of analysis during the 18th century, with a particular focus on two main venues. On the one hand, the book delves into the impact of military and technical teaching (roughly speaking, the teaching of "military engineers") on mathematical education imparted outside universities. On the other hand, it delves into the influence of teaching connected to military and technical needs (e.g., practical geometry, fortification, and architecture) provided in universities. One example in this sense is given by the mathematics instruction offered by the Jesuits.

As a result of the necessity for a highly skilled and efficient military and state officers, the first half of the eighteenth century witnessed a remarkable proliferation of military schools across different European nations, often unrelated to preexisting academic traditions. Unlike the teaching of mathematics in European universities, mathematics in technical and military schools presented more diverse and specialized content, not strictly anchored to the kind of encyclopaedic knowledge sedimented in the traditional *Quadrivium*.

However, the immediate significance of mathematics instruction in the field of artillery was not as obvious as we might think. In his study of French technical education, historian Ken Alder shows that military engineers continued to rely heavily on practical rules and their expertise to teach the art of artillery, even when the formalism of calculus and algebra was available (see, for instance [Alder, 1999, p. 113ff.]). This preference stemmed from the belief that the implicit knowledge passed down by practitioners surpassed the accuracy of existing mathematical

4

models in explaining various phenomena, such as the trajectory of a projectile moving through a resistant medium, for instance, a cannonball shot through the air.

Furthermore, there were debates about which aspects of mathematics were necessary to artillery. Several officers and military men believed that instruction that relied too heavily on abstraction was harmful and useless. Artillery professor B. Forest de Bélidor, whose textbooks were widely popular in French schools and circulated throughout the continent (see for example, M. R. Massa-Esteve's chapter in this volume), argued that engineers should avoid meaningless speculation or endless calculations without purpose. Instead of engaging in the speculations of "snobbish savants," Bélidor proposed that the engineers should focus on a mixture of practical skills and theoretical knowledge ([Alder, 1999, p. 116]).

Since the Renaissance, mathematics applied to physical quantities has been referred to as "mixed mathematics", distinguishing it from the "pure" disciplines of arithmetic and geometry.[1] Broadly speaking, in pre-18th century curricula, mixed mathematics dealt with quantities in conjuction with matter, and included a limited range of phenomena, spanning from mechanics to optics, whose aspects, such as positions, motions and forms could be quantified and measured. However, questions pertaining to the nature and causes of natural phenomena felt outside of the scope of mathematics into the realm of "physics", i.e. natural philosophy.

In the period covered in this book, the advancements in algebra and analysis enabled the mathematical treatment of a broader range of concrete entities and phenomena, referred to as "thick objects" by Ken Alder [2010, p. 71]. These included both natural processes and human activities, with notable examples being projectile motion and war-related artifacts. Although the distinction between "pure" and "mixed" mathematics persisted well into the 18th century, the development of mixed mathematics within artillery and technical schools transformed its original meaning. The new "mixed mathematics" of the Enlightenment aimed to apply mathematical formalism a diverse array of objects,

---

[1][Massa-Esteve, 2011, p. 236]. Tartaglia's work on ballistic, the *Nova Scientia* (1537), was also relevant in this respect. See [Massa-Esteve, 2014].

real-world processes, and phenomena. This expanded scope sought to address the limitations of earlier mixed mathematics, which had been criticized for its inability to explain the majority of natural phenomena. Among the examples discussed in this book, the successes of analysis included investigating applications of algebra and infinitesimal calculus to solve optimization problems, which represented a crucial demand for military and technical professions. These included determining the number of cannonballs in a pile or displacing a mass of earth.[2] or applying mathematics to the dynamics of concrete bodies, such as to calculate the trajectory of a projectile in a resisting medium.

Indeed the expansion of mixed mathematics in technical schools furthered its algebrization and contributed to the primacy of analytical thinking, steering the gradual transformation of "mixed mathematics" into "mathematical physics." This long process is discussed, for instance, in [Massa-Esteve, 2011], and in Massa-Esteve's chapter of this volume.

However, the algebrization of mathematics did not merely impose itself on the basis of its epistemic and cognitive virtues. As we have seen above, these virtues were sometimes the main reason for criticizing the usefulness of mathematics in artillery curricula. "Abstract" disciplines, such as finite and infinitesimal analysis, were often viewed with suspicion when cultivated for their own sake. However, in technical and military curricula, they may have contributed to inculcating a particular social role into artillery officers from the mid to late 18th century, who came to acquire a unique blend of practical skills, theoretical knowledge, and mathematical precision. This mix shaped the ideal image of the "engineer" in the late eighteenth century.[3]

This process happened through the essential role of people who acted as intermediaries, passing on and disseminating mathematical knowledge. To provide just one exemplary case, discussed in this volume, let us mention the mathematician and engineer Benjamin Robins (1706-1751). Despite receiving little attention from historians, Robins was an influential figure in the 18th century. He authored a work called "New Principles of Gunnery" in 1742, which was later translated into German

---

[2]This was a practical problem leading to original developments in mathematics. The problem is mentioned in Patergnani and Lugaresi's chapter in this book.
[3]This thesis is explored, in particular, in [Alder, 1999].

by Euler in 1745. Euler added commentaries that incorporated infinitesimal calculus. Robins' book, through Euler's contributions, became a widely used reference text in artillery schools in Germany, France, and other countries like Spain and Italy, contributing to demonstrate the importance of studying mathematics within technical education (see [Barrow-Green, 2010]). In a similar way, investigating the practices and contributions of other characters marginalized as "dwarfs" in many historic narratives, such as ordinary teachers, administrators, reformers, and textbook authors, can enrich our understanding of how mathematical knowledge was shared and circulated, and how the geography of disciplinary knowledge was modified.

2 Giants and dwarfs as historiographical categories

"Giants" and "dwarfs", "heroes" and "commoners", "luminaries" and "obscure" are metaphors commonly used in narratives of science to contrast great pioneers and innovators and less influential scientists, or those scientists who merely built upon existing knowledge. Are such categories useful in the history of science, especially, history of mathematics? How can they be employed to increase our knowledge of mathematics and its history? These are some of the questions raised by the contributions collected in this volume.

Narratives about the history of 18th century mathematics have often followed the structure of the "great-men narrative", focussing on the lives and work of a small number of canonical individuals considered "giants", "great minds" or "heroes", who made significant contributions to the discipline and towered over less original characters. For example, the account of mathematics in 18th century Europe offered by the renowned historian of mathematics D. Struik begins with a list of mathematicians forming a sort of "intellectual kinship" ([Struik, 1987/1948, p. 163]). Leibniz, Newton, the Bernoulli family, Euler, Lagrange, and a few other French mathematicians shaped, in Struik's narration, that has become quite standard, we find the main features of 18th century mathematics. These are an emphasis on the application of algebra and analysis to domains such as geometry and number theory, on free manipulations and experimentation with symbolic expressions over the rigorous study of the foundations, and the triumph of the "quantifying spirit," namely the idea that mathematics is a language applicable to disparate domains of the study of nature.

This narrative has been subject to scrutiny from various angles in recent years, but precisely those individuals whose achievements may have slipped through the cracks or failed to have a global impact on the discipline deserve our attention. Their significance lies in how they shaped their context, fostered connections among scientists, or played a pivotal role in the dissemination of knowledge, as in the case of teachers and intelligencers. In particular, historical studies of mathematics can play a crucial role in reviving the works of these alleged "dwarfs" by highlighting their contributions to the development of the field and reconstructing their biographies and social contexts. Additionally, shedding light on obscured contributors or groups can help us understand how knowledge is produced and transmitted and how historical canons are formed, including the one that shapes Struik's narration.

One direction of research that will be pursued in this volume is to investigate how certain mathematicians who were once renown as "giants" have become "dwarfs", as they have been forgotten despite having played a non-negligible, or even outstanding public role. This trend follows a recent growing emphasis on rediscovering forgotten personalities in various fields, such as women in mathematics, mathematics teachers, and textbook authors. Emilie du Chatelet (1706-1749) and Maria Gaetana Agnesi (1718-1799) serve as prime examples of remarkable individuals highly acclaimed in their respective times, encompassing all three categories: women, mathematicians, and textbook authors.[4] While this book does not delve into their stories, the recent reassessment of their significance as mathematicians in 18th-century Europe provides a fundamental methodological lesson that we tried to follow in our book too. In the case of Agnesi, scholars such as P. Findlen [2011] and M. Mazzotti [2007] have reexamined archival resources, challenging the critical judgment that relegated Agnesi to the ranks of second-rate, unorigi-

---

[4]On E. du Chatelet, see, for instance, [Hagenbruger, 2011], and the project pursued by the group *History of Women Philosophers and Scientists* (https://historyofwomenphilosophers.org/project/directory-of-women-philosophers/du-chatelet-emilie-1706-1749/).

nal mathematicians of her time. As documented in the aforementioned studies, without a comprehension of Agnesi's biography in her social and intellectual context, one can be fooled into believing that she was not an "original" mathematician. On the contrary, she embodied an original career leading to the acknowledgment by the intellectual world of her time and to a university chair, albeit one she never practically occupied. Moreover, she stood out as a paradigmatic textbook author, representing a model for other female mathematicians.

In addition to recognizing the forgotten giants, historians can also acknowledge groups or individuals who had failed to receive public recognition in the past. As E. Robson and J. Stedall remarked "To limit the history of mathematics to the history of mathematicians is to lose much of the subject's richness" ([Robson and Stedall, 2009, p. 2]). A more comprehensive and richer study of practices should include "cloth weavers, accountants, instrument makers, princes, astrologers, musicians, missionaries, schoolchildren, teachers, theologians, surveyors, builders, and artists". As the various contributions in Stedall and Robson's book show, the categories listed above played a crucial role in the circulation of mathematical knowledge, yet their activities and their traces, whether written or not, have only recently started receiving exploration.

Understanding the role of these often-obscure characters in the production of mathematical knowledge becomes even more crucial when we consider their impact on the so-called "giants" of the field. It is important to recognize that even scientists who claimed to be self-taught or were seen as such did not acquire their expertise out of thin air. These individuals were not solitary figures; rather, they were immersed in a network of influences that contributed to the development and refinement of their mathematical expertise. Their knowledge was shaped by the correspondence networks they were part of, their access to books and journals, and the relationships they formed, including their interactions with fellow mathematicians and practitioners. By delving deeper into the intellectual environment surrounding prominent mathematicians, we gain a richer understanding of how their expertise was constructed and situated.[5]

While the biographical method has been widely used for "giants", it can pose challenges when applied to the study of "dwarfs" who played significant and indispensable roles in the education of the former, for instance, but whose lives are often poorly documented due to limited available sources. To address this issue, one possible approach is to situate their lives and careers within larger groups, such as networks they were part of, more prominent colleagues, friends or correspondents, or the institutions where they may have been employed. A more comprehensive understanding of their contributions can be achieved by examining their connections and associations.

Studying individuals considered "dwarfs" from the perspective of standard histories of science offers more than a means to unravel the biographies of outstanding individuals and shed light on their life and scientific production. It also proves essential in comprehending some of the broad historical phenomena that shaped mathematics in the 18th century and beyond.

One example, which is also a transversal theme of our book, is the process of habituation to analytic formulas and the transformation of algebra (or, as it used to be known in the 17th and 18th centuries, "analysis of finite quantities") and differential and integral calculus (also known as "infinitesimal analysis"), from methods applied to geometry and arithmetic into a new branch of mathematics, simply called "analysis", during the 18th century. In the words of the historian of mathematics Henk Bos:

> Explicit construction as basis for understanding the objects of mathematics was replaced by a trust in the formula, based on a gradually established conviction that the equations of analysis always, explicitly or implicitly, defined an object, and that therefore that object could be accepted as given or as existent. A process of habituation to the world of formulas and equations finally eliminated the demand for a geometri-

---

[5]Among recent examples in the literature, see the case of Herschel examined by Winterburn [2014], Lagrange studied by Borgato and Pepe [1990], and Bolzano investigated by Fuentes Guillén and Crippa [2021].

cal explanation.[6]

The essential steps in this transformation were the contributions of Viète and Descartes, as highlighted, for instance in [Panza, 1997]. What Bos' and Panza's investigations (see also [Bos, 2001]) do not fully clarify is the question about the conditions, both material and social, that made possible the process through which algebraic language imposed itself as the ground and language of mathematics, and analysis became an autonomous discipline at the very heart of modern scientific education.

A general thesis underlying the contributions in this volume is as follows: in order to fully understand the phenomena of habituation and the emergence of new disciplinary fields in mathematics (and, more generally in science), it is necessary to examine how the "great minds" learned algebra and calculus, often from lesser-known teachers and correspondents, how their new ideas circulated and became established, and how mathematical innovations were adopted and adapted by wider communities of mathematicians, often indirectly related to the original sites of their production. As some recent works have demonstrated (for instance [Warwick, 2003], [Ehrhardt, 2010]), the historical development of ideas in mathematics is not solely the work of "giants" but is shaped by a complex interplay of social and intellectual constraints. Even prominent mathematicians go through education and learn from their teachers, who have often been overlooked in the traditional histories of mathematics.

3 Centres and peripheries

In this book, we also examine the concept of algebrization through the lens of "geographies of knowledge" (for a general presentation of this issue, see [Livingstone, 2013]). This perspective encompasses the dissemination of novel theories and methodologies, particularly those associated with the emergence of algebra and calculus. Thus, we will explore the locations where these theories and methods were taught and acquired, as well as the individuals responsible for imparting and acquiring this knowledge. Additionally, we consider the social and political contexts that influenced knowledge production in this field.

---

[6][Bos, 1996, p. 17].

Apart from the above-mentioned [Livingstone, 2013], [Secord, 2004] and the collections edited by Gavroglu [1999] and Blanco and Bruneau [2020] represent reference works for our investigation. In particular, Gavroglu [1999] and Blanco and Bruneau [2020] have emphasized the appropriation of knowledge as an active process, moving away from the idea of transmission as passive reception. These researchers have considered the circulation of knowledge products, such as manuscripts and textbooks, and the relevance of translations for the transfer of knowledge. They also considered travelers' biographies, namely practitioners and scholars who crossed national borders. The circulation of scientific knowledge is now understood as more complex than simply receiving and reproducing it.

The contributions presented in this book align with these trends and investigate the transmission of mathematical knowledge through the circulation of manuscripts, textbooks, their translations, and the itineraries of single scholars and traveling mathematicians such as J. Wendligen (1715-1790), discussed in J. Berenguer's chapter or P. Calbó Caldès (1752-1817), studied by A. Roca Rosell.

A common denominator among the various chapters of the book is the circulation of mathematics from the viewpoint of certain European peripheries, such as Spain and Italy, during the 18th century. These peripheries can be considered "dwarfs" in contrast to "giants" such as centers of knowledge production in the 18th century like France or Prussia. Even if at the time the dichotomy was not stated in these terms, there were undoubtedly places recognized as centers exporting models of academic knowledge. For example, artillery schools in France were funded by Louis XIV and became models to be imitated by other nations, such as Italy (as studied in E. Patergnani and M. G. Lugaresi's contribution) and England. Later, Napoleonic campaigns in Europe disseminated French educational models based on the Polytechnic School.

This book aims to show that the relationship between institutional "dwarfs" and "giants" was not that of simple imitation or passive reception. A noteworthy example of this is the flourishing of artillery schools in various Italian states. Although France stood out as a model, local conditions shaped the organization and syllabi of various military academies. In this regard, it is instructive to compare the place of geometry and analysis in the syllabi of military schools in Turin and Verona.[7] In other local contexts such as Naples, synthetic and analytical methods reflected conflicting social and political undercurrent (see the recent [Mazzotti, 2023]).

Scientific novelties also travelled from centres beyond the alps to communities in Italy and Spain. Cases at point are represented by the circulation of the Leibnizian calculus in Italy, studied by Mazzone and Roero [1997], or the Newtonian calculus and physics (see [Mazzotti, 2019] and, more generally, the whole collection of essays containing that publication). Newton, together with French authors such as Clairaut and La Caille, were also influential in Spain.[8]

Similarly, pedagogical novelties also circulated in 18th century Europe through multi-volume mathematical courses such as Wolff' *Elementa matheseos*, reprinted in numerous editions and abridgements.

Moreover, new mathematical courses were created in peripheral regions by adapting and using materials from other books. Among the examples discussed in this book, Tomàs–Vicent Tosca (1651-1723) composed nine volumes between 1707 and 1715, and Benet Baïls (1730-1797) published 11 volumes between 1779 and 1784.[9] These texts also served as models for other authors.

Another example, untreated in the book but neverteless worth mentioning is Agnesi's *Istituzioni Analitiche* published in 1748 in Milan, at the time part of the Habsburg monarchy. Despite being produced in a place far from the European cultural and economic centers and by a woman (thus by a member of a class of alleged "dwarfs", at least according to a wrong-headed view in common historiography), it was successful throughout Europe, even in translations. In fact, Agnesi's book circulated both in published translations (such as the partial French edition) and unpublished ones, such as Colson's "The Plan of the Lady's System of Analyticks", a two-volume draft translation of the *Istituzioni*.[10]

---

[7]This comparison is addressed in this volume by Patergnani and Lugaresi.

[8]See [Navarro Loidi, 2020] and Navarro Loidi's chapter for this volume. On the influence of Newton in Spain see [Berenguer, 2021].

[9]In fact, the volumes were written in 1772 and published in 1779. Baïls shows a high level of mathematics in his course. See [Martínez-Verdú et al., 2023].

[10]Traces of the circulation of Agnesi's book are visible through discussions of some passages in her books. For instance, the problem in the area of the oblique cone was

In conclusion, by examining different contexts of knowledge production and acquisition, this volume offers several arguments to confirm that the transfer of knowledge is not a straightforward process of information flow from centres to peripheries, since there were networks that overruled the simple centre/periphery division and new knowledge was created in alleged peripheries through the adaptation of new texts to local contexts, or through the reverse circulation of texts from so-called "peripheries" to "centres". We might instead resort to the notion of "dialogue", which involves interaction and interchange: a dialogue between centres and peripheries, between giants and dwarfs. It is in the editors' hope that the question raised and addressed, and the methodologies here deployed could be applied to other case studies, other "peripheries" and other "dwarfs", in Europe and worldwide.

4   The contributions in this volume

The original idea of this book as well as most of its chapters come from a symposium organized by the editors during the latest ICHST-2021 meeting entitled: "Giants and dwarfs in the transformations of mathematics in the XVIII century".[11] The goal of this symposium was to study two interconnected and relevant changes in mathematics that occurred between the middle of 17th to the end of 18th century: the passage of analysis from a method to a discipline regularly taught in colleges and sometimes in universities by the second half of the 18th century, and the transformation of "mixed" mathematics into "mathematical physics" and later "applied" mathematics. In both cases, the perspective of "dwarfs" was chosen as the privileged viewpoint.

The chapter written by F. Gómez García, P. José Herrero Piñeyro, A. Linero Bas, and A. Mellado Romero on Ozanam, as well as E. Dorrego's chapter on Lambert and Legendre study cases of 18th century "giants" who have fallen into oblivion until the recent past. Lambert was deemed as a reference scientific figure by eighteenth century mathematicians and scholars, such as Gauss and Kant, but until a recent

---

the subject of a monograph printed in Spain in 1755 ([Massa-Esteve, 2011, p. 246]), just a few years after the publication of Agnesi's *Istituzioni* (1748).

[11]See: https://www.ichst2021.org.

reappraisal in the literature, it was little known even among specialists. The reasons for Lambert's neglect are partly explained by José Ferreiros' foreword in [Dorrego López, Fuentes Guillén, 2023]. According to Ferreiros' argument, Lambert was a character imbued with the spirit of Enlightenment:

> He was no specialist, but rather the opposite: a philosopher as much as a scientist, he contributed to all the sciences of his time; while active in the Academies of Munich and Berlin, he contributed to all the different «classes» or areas of work. It has been said, that, for bad and good, Lambert was the perfect example of the eighteenth-century erudite, who wrote about God and the world, about all possible topics: mathematics, experimental science, philosophy, languages, and history.[12]

Ozanam represents a similar case of a scholar considered as a second-order mathematician, at least until the recent past.[13] Did his contemporaries also share this judgment? The chapter published in this book offers arguments to consider the opposite. In the late 17th century, Ozanam was a well-known author who enjoyed the protection of important patrons, he was esteemed by his peers and his students, and received public praises in journals. Leibniz even considered Ozanam worthy of being part of the circle of expert mathematicians, as he was one of the most skilled and experienced in performing ordinary calculations effortlessly.

The reputation of Ozanam had changed in the space of a few generations. In his famous *Histoire des mathématiques*, E. Montucla did not spare criticism when he remarked on how Ozanam had to curb to the whims of his public, which distracted him from serious mathematical contributions (in [Schaaf, 1970-1991]). Just as Lambert, Ozanam embodied the character of a "polymath" distant from the idea of the

---

[12][Dorrego López, Fuentes Guillén, 2023, p. viii].

[13]According to the *Dictionary of scientific biographies*: "By almost any criterion Ozanam cannot be regarded as a first-rate mathematician, even of his own time" ([Schaaf, 1970-1991]).

"mathematician as a specialist" that would gain prominence and acceptance during the next century.[14]

The contributions of Berenguer, Navarro Loidi, Massa-Esteve, Roca Rosell, and Patergnani and Lugaresi shed light on the biographies of teachers, reformers, and officers by examining the academic context in which they worked. This context was, for instance, that of technical education in private colleges analyzed by Roca Rosell, that of military academies studied by Navarro Loidi, Massa-Esteve, and Patergnani and Lugaresi, or that of Jesuit scientists studied by Navarro Loidi and Berenguer.

Mateo Calabro and Pedro Lucuce, professors at the Royal Military Academy of Barcelona, Tomas Morla and Pietro Giannini, teachers at the Royal Military College of Knights Cadets of Segovia, are the key figures analyzed by Navarro Loidi as instrumental for the establishment of a modern technical curriculum in Spain during the second half of the eighteenth century. During that period, algebra and calculus became essential components of the curricula in technical schools. An intriguing aspect of the story addressed by Navarro Loidi is that these disciplines were not accepted into teaching without opposition. Among the examples studied by Navarro Loidi, the gunner and teacher Tomás Morla recognized, in his Treatise on Artillery (1784-86), that mere practice was "blind and servile" when "divested of principles and theory." However, Morla opposed the excessive use of mathematics in military schools, arguing that military men were not "astronomers" and did not require the same type of theoretical instruction as students of "pure mathematics." Morla's target was Pietro Giannini's course written for the Military Academy of Segovia. The course was deemed too theoretical, and its author, Giannini, was accused of being a "mathematician" but not a "gunner". Finally, the combination of scientific progress, political support, and institutional leadership helped overcome resistance to the inclusion of algebra and calculus in military education in Spain during the 18th century.

By studying the beginnings of the Royal Military Academy of Mathe-

---

[14]Such a change in the image of the mathematician is examined by A. Alexander ([2006]). Broadly speaking, Alexander relates changes in the ways mathematicians were perceived as public figures to changes in mathematical practices that occurred at the turn of the nineteenth century.

matics of Barcelona (1720) using original sources, M. Rosa Massa-Esteve sheds light on the key transformations of mathematics between the 17th and 18th century: the transition of analysis from being a method to becoming a discipline regularly taught in colleges and universities, the transformation of "mixed" mathematics into "physic mathematics", and the advancements in the classification of mathematical disciplines during this period. In addition, the story of Jorge Prospero Verboom, creator of the Spanish Corps of Military Engineers, which was officially approved in 1711, stands out as yet another example of a "dwarf" who, despite being downplayed in the history of mathematics, were important for the aforementioned transformations. Another character explored in Massa-Esteves chapter is Pedro Lucuce, who was appointed as director and mathematics teacher at the Royal Military Academy of Mathematics of Barcelona from 1738 to 1756 and from 1760 to 1779.[15] Such as the case of other mathematics teachers populating this volume, Lucuce prepared his courses selecting the content from various works circulating in Europe and adapting them to his audience, looking for a balance between innovation and adherence to classical authorities.[16]

The objections raised against the introduction of abstract and higher mathematics are also discussed in Patergnani and Lugaresi's chapter on the history of Italian artillery schools. The authors mention the course offeed by Lagrange at Turin's artillery school in 1755 and how his approach was perceived as too abstract and overly challenging for future officers' needs. Lagrange's course entered deep discussions on infinitesimal calculus and analytic geometry, which were considered irrelevant for military education.

The chapter discussed by Lugaresi and Patergnani charts the evolution of technical teaching institutions in Italy from the early 18th century to the second half of the 19th century, and shows that the presence of military schools became a crucial factor in understanding the circulation of mathematics during this period. Italy's military schools were fashioned after the French system, with the Napoleonic era playing a particularly

---

[15] In the years from 1757 to 1760, Lucuce was director of the Royal Military Society of Mathematics for elaborating and printing a Mathematical Course. Eventually, the project of this Society failed. See [Blanco and Puig-Pla, 2020].

[16] Lucuce's course was used until 1803, when the Royal Military Academy of Mathematics of Barcelona was closed.

influential role in shaping the country's own military education models. This period witnessed a geopolitical transformation in Italy, triggered by Napoleon's actions, resulting in the removal of rulers from the *Ancien Régime* and the establishment of "sister republics" modelled after French institutions. Among these institutions, the Royal Polytechnic and Military School of Naples adopted the French model. Similarly, the school in Pavia also experienced the impact of French influence during the Napoleonic period. One noteworthy consequence of this influence was the dissemination of educational resources, textbooks, and teaching methods, that extended beyond military education and made their way into broader technical education. This influence continued to shape engineers training at universities in the following century.

The subject of Berenguer's article is another alleged "dwarf", the bohemian teacher Johannes Wendlingen. Wendlingen's career sheds light on the efficient jesuit academic networks that facilitated the mobility of teachers between various European and non-European countries.

In mid-eighteenth century Spain, the Society of Jesus was one of the intellectual groups in the service of the Crown in the process of modernizing the country. Science was an essential factor in promoting a renewal movement aimed at consolidating the Monarchy, and the Society of Jesus became one of the institutions capable of implementing it and guiding the establishment of the new science.

Berenguer's chapter explores Johannes Wendlingen's main achievements as a mathematician, astronomer and cosmographer in the service of the Spanish Crown. Wendlingen was entrusted with the task of writing a complete mathematics course to be employed as a textbook at the Imperial College in Madrid. Of particular significance is the section of the treatise dedicated to differential calculus, which reveals that Wendlingen was instrumental in its introduction to Spain - a field that was relatively unfamiliar in the country at the time. Berenguer's chapter also provides a detailed comparative analysis showing how Wolff's *Elementa* was used as a guide by Wendligen to teach arithmetic, practical geometry, and infinitesimal calculus. Berenguer's discussion can be taken as an example illustrating how the reception of a text is determined by both local contexts and the purposes for which it is reused. This can vary among mathematical communities and individuals. In

Berenguer's case study, there was a need to adapt Wolff's exercises in geometry to meet the practical needs of topographers in Spain.

Finally, the lesser-known teacher and textbook author Calbó Caldés is studied in A. Roca Rosell 's chapter. Pasqual Calbó Caldés was an artist-scientist from Minorca who taught a course in mathematics in the late 18th century. The manuscript of Calbó's course, written in Minorcan Catalan and analyzed by Roca-Rosell, provides us with valuable insight into the history of private technical education and the role of mathematics in it during the Enlightenment.

Furthermore, Calbó was a living example of how knowledge circulated not only through books but also through people, spending nine years among Venice, Rome, and Vienna. In these cities, he trained as an artist, which may have influenced his own activity as a mathematics teacher. Calbó's manuscript contains a wealth of knowledge on pure and mixed mathematics, showing an interest in experimental physics, sundials, perspective, architecture, and shipbuilding.

All in all, Roca Rosell's chapter is an example of the importance of researching non-academic environments, such as private teaching settings and "workshop cultures," namely the private environments where soon-to-be engineers and architects received their training, often via the contact of experts in these professions, before the ultimate setting up of dedicated schools in the 19th century.

During the preparation of this work the authors used paperpal.com and chat.openai.com in order to provide linguistic revisions. After using these services, the author(s) reviewed and edited the content as needed and take full responsibility for the content of the publication.

References

[Alder, 1999] K. Alder. French Engineers Become Professionals; or, How Meritocracy Made Knowledge Objective. In Clark, William, Jan Golinski, and Simon Schaffer, editors. *The Sciences in Enlightened Europe*, pages 94–126. University of Chicago Press, Chicago, 1999.

[Alder, 2010] K. Alder. *Engineering the Revolution: Arms and Enlightenment in France, 1763-1815*. The University of Chicago Press, Chicago, London, 2010.

[Alexander, 2006] A. R. Alexander, Tragic Mathematics: Romantic Narratives and the Refounding of Mathematics in the Early Nineteenth Century. *Isis*, 97(4): 714–26, 2006. https://doi.org/10.1086/509952.

[Barrow-Green, 2010] J. Barrow-Green. Euler as an Educator. *BSHM Bulletin: Journal of the British Society for the History of Mathematics*, 25(1): 10–22, 2010. https://doi.org/10.1080/17498430903321141.

[Berenguer, 2021] J. Berenguer. Introducing differential calculus in Spain: The fluxion of the product and the quadrature of curves by Tomàs Cerdà. *British Journal for the History of Mathematics*, 36 (1): 23–49, 2021.

[Blanco and Bruneau, 2020] M. Blanco, O. Bruneau. Introduction. Les mathématiques dans les écoles militaires. *Philosophia Scientiae*, 24(1):5–11, 2020.

[Blanco and Puig-Pla, 2020] M. Blanco and C. Puig-Pla. The Role of Mathematics in Spanish Military Education in the 1750's: Two Transient Cases. *Philosophia Scientiae*, 24(1): 97-113, 2020.

[Borgato and Pepe, 1990] M. T. Borgato, L. Pepe. *Lagrange. Appunti per una biografia scientifica*. La Rosa, Torino, 1990.

[Bos, 1996] H. Bos. Johann Bernoulli on Exponential Curves. *Nieuw Archief Voor Wiskunde*, 14(1): 1–19, 1996.

[Bos, 2001] H. Bos. *Redefining Geometrical Exactness: Descartes' Transformation of the Early Modern Concept of Construction*. Springer Verlag, New York/Berlin/Heidelberg, 2001.

[Dorrego López, Fuentes Guillén, 2023] E. Dorrego López, E. Fuentes Guillén. *Irrationality, Transcendence and the Circle-Squaring Problem: An Annotated Translation of J. H. Lambert's Vorläufige Kenntnisse and Mémoire*. Vol. 58. Logic, Epistemology, and the Unity of Science. Cham: Springer International Publishing, 2023.

[Ehrhardt, 2010] C. Ehrhardt. Histoire Sociale Des Mathématiques. *Revue de Synthèse*, 131(4): 489–93, 2010. https://doi.org/10.1007/s11873-010-0131-2.

[Findlen, 2011] P. Findlen. Calculations of Faith: Mathematics, Philosophy, and Sanctity in 18th-Century Italy (New Work on Maria Gaetana Agnesi). *Historia Mathematica*, 38(2): 248–291, 2011.

[Fuentes Guillén and Crippa, 2021] E. Fuentes Guillén, D. Crippa. The 1804 Examination for the Chair of Elementary Mathematics at the University of Prague. *Historia Mathematica*, 57: 24-54e.18, 2021. `https://doi.org/10.1016/j.hm.2021.07.001`.

[Hagenbruger, 2011] R. Hagengruber. *Emilie du Châtelet between Leibniz and Newton*. Springer, Dordrecht, 2011.

[Gavroglu (ed.)), 1999] K. Gavroglu, editor. *The Sciences in the European Periphery During the Enlightenment*. Springer Netherlands, Dordrecht, 1999.

[Klein, 1976/1936] J. Klein. *Greek Mathematical Thought and the Origin of Algebra*. M.I.T Press, Cambridge, 1976. Translated from the original German edition *Die griechische Logistik und die Enstehung der Algebra*, in *Quellen und Studien zur Geschichte der Mathematik, Astronomie, Physik*, Abteilung B: Studien, Vol. 3, Fasc. 1 (Berlin, 1934), pages 18–105 (part I); fasc. 2 (1936), pages 122–235 (Part II).

[Livingstone, 2013] D. N. Livingstone. *Putting Science in Its Place: Geographies of Scientific Knowledge*. *Science.Culture*. University of Chicago Press, Chicago, 2013.

[Mahoney, 1980] M. S. Mahoney. The Beginnings of Algebraic Thought in the Seventeenth Century. In *Descartes, Philosophy, Mathematics and Physics*, Harvester Readings Hist. Sci. Philos., No. 1:141–55. The Harvester Press and Barnes & Noble Books, Brighton and New Jersey, 1980.

[Mancosu, 1989] P. Mancosu The Metaphysics of the Calculus: A Foundational Debate in the Paris Academy of Sciences, 1700-1706. *Historia Mathematica*, 16: 224-248, 1989

[Mancosu, 1996] P. Mancosu. *Philosophy of Mathematics and Mathematical Practice in the Seventeenth Century*. Oxford University Press, New York, 1996.

[Maronne and Panza, 2014] S. Maronne, M. Panza. Euler, Reader of Newton: Mechanics and Algebraic Analysis. *Advances in Historical Studies*, 3(1): 12–21, 2014.

[Martínez-Verdú et al., 2023] D. Martínez-Verdú, M. R. Massa-Esteve, and A. Linero-Bas. Infinite Analytical Procedures for the Computation of Logarithms in Works by Benito Bails (1731–1797). *British Journal for the History of Mathematics*, online first: 1—34, 2023. https://doi.org/10.1080/26375451.2023.2186648.

[Massa-Esteve, 2001] M. R. Massa-Esteve. Las relaciones entre el álgebra y la geometría en el siglo XVII. *Llull*, 24: 705–725, 2001.

[Massa-Esteve, 2006] M. R. Massa-Esteve. *L'Algebrització de les Matemàtiques. Pietro Mengoli (1625-1686)*. Societat Catalana d'Història de la Ciència i de la Tècnica, Barcelona, 2006.

[Massa-Esteve, 2011] M. R. Massa-Esteve, A. Roca-Rosell, and C. Puig-Pla. Mixed Mathematics in Engineering Education in Spain: Pedro Lucuce's Course at the Barcelona Royal Military Academy of Mathematics in the Eighteenth Century.*Engineering Studies* 3(3): 233–53, 2011.

[Massa-Esteve, 2014] M. R. Massa-Esteve. Historical activities in the mathematic's classroom: Tartaglia's Nova Scientia (1537). *Teaching Innovations*, 27(3):114–126, 2014.

[Massa-Esteve, 2020] M. R. Massa-Esteve. The Algebraization of Mathematics: Using Original Sources for Learning Mathematics. *Inovacije u Nastavi* 33(1): 21–35, 2020. https://doi.org/10.5937/inovacije2001021E.

[Mazzone and Roero, 1997] S. Mazzone, C. S. Roero, *Jacob Hermann and the diffusion of the Leibnizian Calculus in Italy*. Biblioteca di Nuncius: Studi e testi, 26. Olschki, Firenze, 1997.

[Mazzotti, 2007] M. Mazzotti. *The World of Maria Gaetana Agnesi, Mathematician of God*. Johns Hopkins Studies in the History of Mathematics. Johns Hopkins University Press, Baltimore, 2007.

[Mazzotti, 2019] M. Mazzotti. Newton in Italy. In H. Pulte and S. Mandelbrote (eds.), *The Reception of Newton in Europe*, 3 vols, pages 159–178. Bloomsbury Publishing, 2019.

[Mazzotti, 2023] M. Mazzotti. *Reactionary Mathematics: A Genealogy of Purity*. The University of Chicago Press, Chicago, 2023.

[Morton Briggs, 1970-1991] J. Morton Briggs. Robins, Benjamin. In *Complete Dictionary of Scientific Biography*, Scribner, New York, 1970-1991. Consulted online at: Encyclopedia.com. Accessed 26 May 2023.

[Navarro Loidi, 2020] J. Navarro Loidi. Foreign Influence and the Mathematics Education at the Spanish College of Artillery (1764-1842).*Philosophia Scientae*, 24(1): 115–36, 2020. https://doi.org/10.4000/philosophiascientiae.2192.

[Panza, 1997] M. Panza. Classical sources for the concepts of analysis and synthesis, in M. Panza and M. Otte (editors), *Analysis and synthesis in mathematics. History and philosophy*, pages 365–414. Boston Studies in the Philosophy of Science 196. Kluwer Academic Publishers, Dordrecht, 1997.

[Robson and Stedall, 2009] E. Robson, J. A. Stedall, editors. *The Oxford Handbook of the History of Mathematics*. Oxford University Press, Oxford, New York, 2009.

[Schaaf, 1970-1991] W. Schaaf. Ozanam, Jacques. In *Complete Dictionary of*

Scientific Biography, Scribner, New York, 1970-1991. Consulted online at: Encyclopedia.com. Accessed 26 May 2023.

[Secord, 2004] J. A. Secord. Knowledge in Transit. *Isis* 95(4): 654–72, 2004. https://doi.org/10.1086/430657.

[Struik, 1987/1948] D. J. Struik. *A Concise History of Mathematics*. Dover, New York, 1987 (originally published in 1948).

[Warwick, 2003] A. Warwick. *Masters of Theory: Cambridge and the Rise of Mathematical Physics*. The University of Chicago Press, Chicago, 2003.

[Winterburn, 2014] E. Winterburn. Philomaths, Herschel, and the Myth of the Self-Taught Man. *Notes and Records of the Royal Society of London*, 68(3): 207–225, 2014. https://doi.org/10.1098/rsnr.2014.0027.

[Wolff, 1717] C. Wolff. *Elementa Matheseos Universae*, volume 1. Renger, Halle an der Saale, 1717.

Jacques Ozanam: Master of the Diffusion and Algebrization of Mathematics

Francisco Gómez García
IES Ramón y Cajal Murcia, Spain
pacogomez11@gmail.com

Pedro José Herrero Piñeyro
IES Beniaján, Murcia, Spain
pedrojose.herrero@murciaeduca.es

Antonio Linero Bas
Departamento de Matemáticas, Universidad de Murcia, Spain
lineroba@um.es

Antonio Mellado Romero
IES Licenciado Francisco Cascales Murcia, Spain
antmero@gmail.com

Abstract

Jacques Ozanam (1640-1718) developed his mathematical work between the seventeenth and eighteenth centuries, at a time of profound change in the way of thinking and doing mathematics. The new algebraic language, initiated by François Viète (1540-1603) and, a little later, by René Descartes (1596-1650), brought about this profound change. Ozanam is a clear exponent of a mathematician who adopted this new language and new style of thinking. He published a large number of works that addressed all well-known aspects of mathematics. His works were widely disseminated and known, as evidenced by the fact that they were reprinted a great number of times, even after his death, and they were translated into English. In addition, we will show, by analyzing some of his works, how he was an extraordinary and giant mathematician who was praised by other distinguished mathematicians such as Gottfried W. Leibniz (1646-1716) or Jacques De Billy (1602-1679).

1 Introduction

The seventeenth century was born, as far as mathematics are concerned, when the ink in the book *In Artem Analyticen Isagoge* [Viète, 1591] was still fresh on the paper. This work was written by the French mathematician François Viète (1540-1603), who established connections between the new algebra, which he was initiating, and the solving of geometric problems. Viète's work was a great novelty. He introduced the use of symbols -letters- in solving problems, to represent both unknown and known quantities, which made it possible to search for solutions to equations, and therefore solutions to problems, in a general way.

We can place in Viète's work the beginning of a profound change in the way of doing and thinking about mathematics. The new algebraic methods prompted the evolution from an eminently geometric thought to another, more algebraic form of mathematical thought, which meant a paradigm shift in mathematical work. This process, developed mainly throughout the seventeenth century, is known as the "algebrization process", and it allowed a rapid and momentous evolution of mathematics with the appearance of analytical geometry and infinitesimal calculus. In Mahoney's words,[1] "the most important and basic achievement of mathematics at the time, to wit, the transition from the geometric mode of thought to the algebraic".

The new algebraic language was gradually adopted to address mathematical problems and to write mathematical treatises, as it happened in the prominent cases of René Descartes (1596-1650) and Pierre de Fermat (1601-1665). In 1637, Descartes published *La Géométrie* as part of his treatise *Discours de la méthode* [Descartes, 1637], which became the most influential work on the evolution of the relationship between

---

[1]Mahoney [Mahoney, 1980, p. 141] has studied the impact, not without tension, of this new way of algebraic thinking: "Close examination of the works of leading mathematicians of the seventeenth century often reveals a certain tension between two modes of mathematical thought: an old, traditional, geometric mode and a new, in many ways revolutionary, algebraic mode", and he gives it great importance in the process of change in mathematics throughout the seventeenth century: "In the light of the brilliant mathematical achievements of the later seventeenth century, in particular the infinitesimal calculus, there is a risk of overlooking the most important and basic achievement of mathematics at the time, to wit, the transition from the geometric mode of thought to the algebraic".

geometry and the new algebra.

In the years following the publication of *La Géométrie*, this new mode of algebraic thinking prevailed, allowing the rapid evolution of mathematics throughout the 17th century.[2] Classical treatises updated with new languages proliferated. Among them, Euclid's *Elements* stands out.

However, alongside the production of the well-known giants of mathematics such as Viète, Descartes, Isaac Newton (1643-1727) or Gottfried W. Leibniz (1646-1716), there were numerous mathematicians who wrote courses, treatises, or monographs that were important and fundamental both for the evolution of mathematics and for its teaching and dissemination, including the new and powerful algebraic language. Without the contribution of Pierre Hérigone (1580-1643),[3] Claude Millet Deschales (1621-1678), Gaspar Bachet (1581-1638), Pietro Mengoli (1626-1686) and so many others, the growth, teaching, and development of mathematics would have been impossible.

Among all these "dwarfs", Jacques Ozanam (1640-1718) stands out. He was a prolific French mathematician who, in addition to publishing numerous works on mathematics,[4] was able to generalize resolutions to numerous problems that he introduced in some of his treatises; to contribute to the consolidation and development of new algebraic methods; and to integrate and relate those methods to the solution of geometric problems, all with great skill and elegance. In addition, his courses and mathematical treatises dedicated to the teaching of mathematics are noteworthy. They have a very remarkable didactic clarity, as shown in the reissues of a good number of them. For example, the *Cours de mathématique* [Ozanam, 1693] was reissued in 1697, 1699; and, in 1712, an edition was translated into English. Another example was the *Récréations mathématiques et physiques* [Ozanam, 1694], a recreational

---

[2]For more information on the process of algebrization see, among others, [Bos, 2001], [Mancosu, 1996] or [Massa-Esteve, 2001], [Massa-Esteve, 2012].

[3]Hérigone wrote a Mathematical Course in six volumes, edited between 1634 and 1642. In [Mellado Romero, 2022], we find an in-depth study on the influence of this Course on the algebrization process of mathematics.

[4]C. Càndito [Càndito, 2016] references 25 works by Jacques Ozanam and takes a look at the content of the main works published by him. She also offers a possible explanation of Ozanam's mathematical motivations and interests.

work (as its title indicates) addressed to a mathematics-loving public, which had, until 1844, at least 17 reissues, several of them translated to English.

In this chapter, we will focus on the relationship between Ozanam and the new algebraic methods, their didactics, and their development. We will also focus on how Ozanam synthesizes algebra with general problem-solving and geometric conclusions and applications in three of his works, such as the *Cours de mathématique*; his treatise on algebra entitled *Nouveaux elements d'Algebre* [Ozanam, 1702]; and *Les six livres de l'arithmetique de Diophante d'Alexandrie augmentez & reduits a la specieuse* [Ozanam, 1674?], an unpublished handwritten work by Ozanam that was lost for centuries, was found by Jean Cassinet (1928-1999),[5] and has been recently studied in [Gómez et al., 2021]. To show the excellence of Ozanam's work, we will analyze some of the problems included in his *Diophantus*, highlighting his mastery of algebraic language as well as the versatility of his thinking and how he was able to relate, in the same problem, different domains of mathematics: numbers, arithmetic, algebra, and geometry.

2   Jacques Ozanam

2.1   Biographical notes

Ozanam was born in 1640 in Sainte-Olive (Ain), France, into a wealthy family of old Jewish tradition that had converted to Catholicism. Nevertheless, he did not inherit anything from the family estate, which went, according to the customs of the time, to his older brother. It was probably this circumstance, and the desire to ensure his future, that led his father to direct him to the ecclesiastical studies, that Ozanam began with the Jesuits. However, what Ozanam really liked was mathematics, and after the death of his father, he abandoned religious studies to devote himself fully to studying the former science. The abandonment of the Jesuits was by no means traumatic, as he maintained a good relationship with them when it came to mathematics. He was a disciple of

---

[5] Jean Cassinet was a French mathematician and mathematics historian, professor at the University of Toulouse. See [Cassinet, 1986].

the Jesuit Claude F. Millet Deschales (1621-1678) and it is possible he was also a disciple of the Jesuit Father Jacques De Billy (1602-1679). Both of them were experts and well-known mathematicians at the time, and he maintained frequent correspondence with both of them.

Ozanam devoted himself to studying and teaching mathematics classes in Lyon. This provided him with a possibly modest livelihood. An event that occurred in Lyon radically changed his life. According to Bernard le Bovier de Fontenelle,[6] (1657-1757) [Fontenelle, 1719] Ozanam taught mathematics in Lyon to two foreigners who, when the time came, told him that they could not return to Paris because they had not received the letters with the necessary credit. Ozanam selflessly loaned them money without even demanding a receipt. These two characters must have been related to the father of the Chancelier of Paris, the nobleman D'Aguesseau,[7] who, seeing Ozanam's generosity, ordered for him to be brought to Paris, promising to help him settle in and make his mathematical abilities known.

From then on, Ozanam would reside in Paris where he was well-known as a mathematician and maintained a relationship with most of the leading mathematicians of the time.

He married Françoise Frevost in 1674 and became widowed in 1701. Apparently, the death of his wife plunged him into a crisis from which he recovered slowly. Finally, Ozanam died in Paris in 1718. As a curiosity, notice that, although Fontenelle, in his "Eloge" placed Ozanam's death in 1717, he actually died in 1718.[8]

2.2 The mathematician Ozanam

As we shall see, the success of Ozanam's mathematical work resembles that of "dwarf" David, rather than a winner Goliath, but he stands

---

[6]Fontenelle was perpetual secretary of the *Royale Académie des Sciences* from 1697 and, after Ozanam's death, published in 1719 a panegyric with the title of *Eloge de M. Ozanam* [Fontenelle, 1719], in which he collected biographical notes on Jacques Ozanam, as well as his main works and the most relevant aspects of his life.

[7]Henri François D'Aguesseau (1668-1751) was Chancelier of France, a high position appointed directly by the king whose role was the administration of justice in France.

[8]More details on Ozanam's biography in [Fontenelle, 1719], [Schaaf, 1990] or [Càndito, 2016].

next to the giants and does not detract in height. Ozanam published many works: specialized treatises, both in pure and applied mathematics, mathematics courses, a mathematical dictionary and his well-known mathematical recreations; in addition to some articles in *Le Journal des Sçavans*.[9]

His works were, in large numbers, reissued or reprinted and some translated into English, even after his death. As an example, let us just comment on some of them. The *Traité de gnomonique* [Ozanam, 1673] was published in 1673 and reissued in 1685 and 1730; *La Géométrie pratique*, was first published in 1684 and reprinted in 1689, 1693, 1711, 1736 and 1762. Treatises on applied mathematics such as the treatise on land measurement *Traité de l'arpentage* [Ozanam, 1699], first published in 1699 and republished in 1725, 1747, 1758, 1779 and 1803; or the treatise, also on land measurement, *L'Usage du Compas de Proportion* [Ozanam, 1688], first published in 1688 and reprinted five more times until 1794.

We have already cited his *Cours de Mathématique* (see Figure 1) [Ozanam, 1693] which consists of 5 volumes and which, in addition to being reissued twice in French, was published in London in an English version in 1712 [Ozanam, 1712]. And the one that undoubtedly stands out in terms of dissemination is *Récréations mathématiques et physiques* [Ozanam, 1694] which since its first edition in 1694 has been reissued nine times in Paris, and three times in Amsterdam. It has also been translated into English with one edition in Dublin and five in London, the last one in 1851 under the title *Recreations in science and natural philosophy* [Ozanam, 1851].[10]

As we have already stated, it is clear that despite this small sample, his figure as a rigorous mathematician, who shows precise writings when exposing this science, acquires an extraordinary dimension. Moreover, a generation was formed with his treatises and his courses.

---

[9] *Le Journal des Sçavans* was the first scientific journal published, although it collected various types of work, from reviews of published books to works on medicine, nature, history or mathematics, among many others. The first number was published in Paris, on January 5, 1665. 259 years of this publication are available from the Bibliothèque Nationale de France; visit `https://gallica.bnf.fr`.

[10] A more detailed, though not complete, account of the works of Ozanam can be found in [Càndito, 2016] or in [Pelay, 2011].

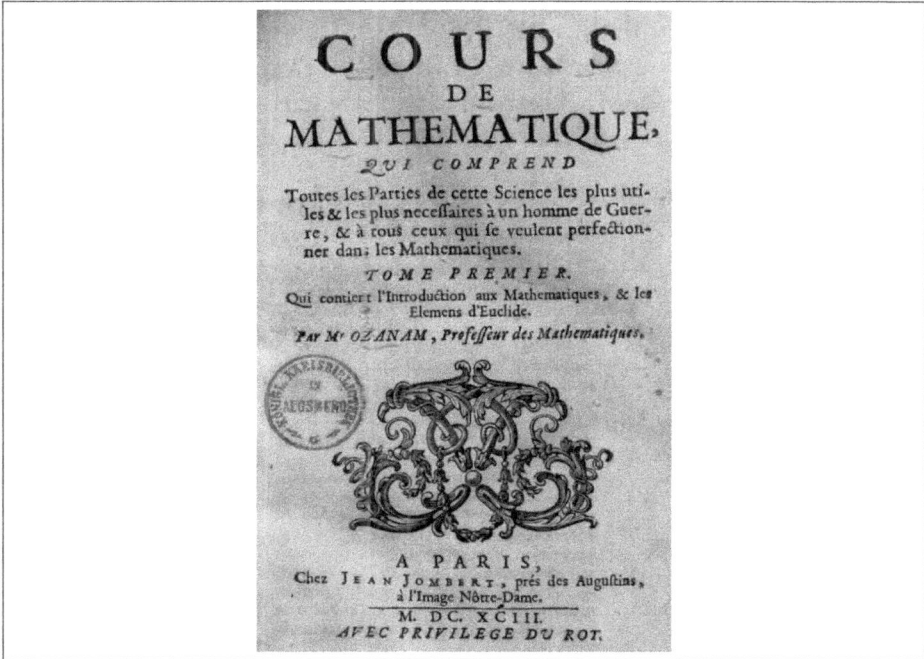

Figure 1: Title page of the first volume of Ozanam's *Cours de Mathématique* (`https://mdz-nbn-resolving.de/details:bsb11267609`).

His mathematical talent, however, stems not only from the volume of his works and the numerous editions or translations that have occurred, but also from the scientific quality recognized by the best mathematicians of the time. Without going any further, a mathematician as prominent as Leibniz wrote about the recent publication of Ozanam's treatise on Algebra [Ozanam, 1702] in the number corresponding of the *Journal des Savants* to Monday, June 11, 1703 [Leibniz, 1703], calling attention to its quality and originality:

> Mr. Ozanam's Algebra, which I have just received, seems much better to me than most of the ones we have seen for some time, which only copy Descartes and his commentators. I am glad that he revives some of the precepts of Viète, inventor of the specious, which deserved not to be forgotten. There are also some very helpful guidelines in the fashion of

31

RÉMARQUE DE M. LEIBNITZ., SUR UN ENDROIT
des nouveaux Elemens d'Algebre de M. Ozanam.

L'Algebre de M. Ozanam, que je viens de recevoir, me paroift bien meilleure que la plufpart de celles qu'on a vûës

depuis quelque temps, qui ne font que copier Defcartes & fes Commentateurs. Je fuis bien aife qu'il faffe revivre une partie des preceptes de Viete, inventeur de la *Specieufe*, qui meritoient de n'eftre point oubliez. On y trouve de plus quelques adreffes tres utiles dans les problemes à la mode de Diophante.

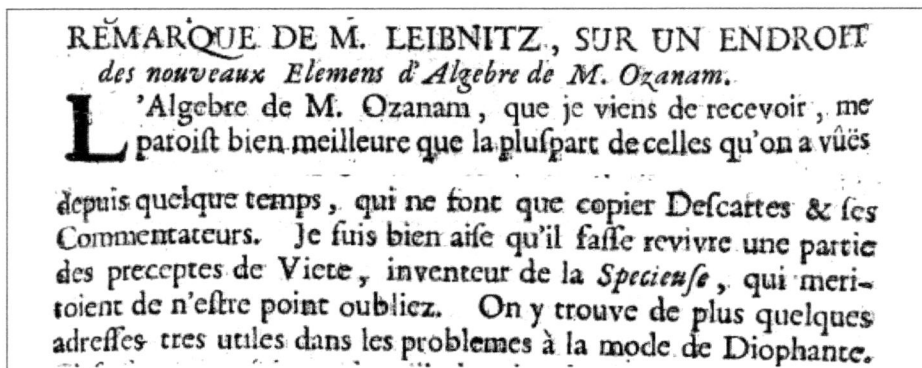

Figure 2: Leibniz, *Le Journal des Savants*, 1703, about *Nouveaux elements d'Algebre* de Ozanam (https://www.digitale-sammlungen.de/en/view/bsb10501916?page=368,369).

Diophantus.[11]

However, not all comments were totally pleasant. In the same issue of *Le Journal des Savants*, precisely at the end of Leibniz's note on p. 364, another anonymous note is included. It is relevant to mention here that, at the time, the editor in charge of mathematics in *Le Journal des Savants* was Fontenelle, see [Sergescu, 1936]. Such a note, with the title *Quelques autres Remarques sur le même Traité d'Algebre à l'occasion de celle de M. Leibnitz*,[12] commented, in a certain critical tone, on Leibniz's own words. For example, with regard to Viète, the author of the note goes on to say that recovering Viète's precepts entails the use of Greek terms that, at that time, were little used by algebraists ([Journal, 1703, pp. 364– 365]). Which does not prevent him from saying a little later, "A few preserved Greek words do not prevent Mr. Ozanam from being

---

[11]"L'Algebre de M. Ozanam, que je viens de recevoir, me paroist bien meilleure que la pluspart de celles qu'on a vûës depuis quelque temps, qui ne font que copier Descartes & ses Commentateurs. Je suis bien aise qu'il fasse revivre une partie des preceptes de Viete, inventeur de la *Specieuse*, qui meritoient de n'estre point oubliez. On y trouve de plus quelques adresses tres utiles dans les problemes à la mode de Diophante", [Leibniz, 1703] pp. 362-363. See Figure 2. Unless otherwise indicated, translations are those of the authors.

[12]*Other notes on the same Treatise on Algebra on the occasion of Mr. Leibniz's.*

clear, intelligible and methodical".[13]

After Ozanam's death, Fontenelle published a panegyric in 1719 with the title *Eloge de M. Ozanam* [Fontenelle, 1719], in which he collected biographical notes on Jacques Ozanam and, among other things, said of his work as a mathematician:

> He wrote with great ease, even on difficult subjects. His first draft was also the final version, never erasures or corrections, and the Printers took advantage of the clarity of his Manuscripts. He sometimes solved problems by going down the street, sometimes even, it is said, asleep, and then, when he woke up, he would take something to write them down; since memory, an almost irreconcilable enemy of judgment, did not dominate in him.[14]

Fontenelle also made a list of what, in his opinion, were Ozanam's main works: *Dictionaire de Mathematiques*, *Cours de Mathematique*, *Traité d'Algebre* or *Sections Coniques*, *Récréations Mathématiques*, and an unpublished handwritten work titled *Diophantus*,[15] part of the library of M. le Chancelier.[16]

Precisely this handwritten work was worthy of praise by other prominent figures of the time. Thus, another Frenchman, Jean-Étienne Montucla (1725-1799), addressed the classic works, among them the *Arithmetic* of Diophantus of Alexandria, in volume one of the first edition of his history of mathematics [Montucla, 1758]. Montucla wrote about

---

[13]"Quelques mots grecs conservez, n'empêchent pas que M. Ozanam ne soit clair, intelligible, methodique", p. 364 [Journal, 1703].

[14]"Il composoit avec une extrême facilité, quoi-que sur des sujets si difficiles. Sa premiere façon étoit la derniere, jamais de ratures ni de corrections, & les Imprimeurs se loüoient fort de la netteté de ses Manuscrits. Quelquefois il resolvoit des Problêmes embarassés en allant par les rües, quelquefois même, dit-on, en dormant, & alors il se faisoit apporter promptement à son réveil de quoi les écrire, car la memoire, ennemie presque irreconcilable du jugement, ne dominoit pas en lui", pp. 89-90 of [Fontenelle, 1719].

[15]This is the handwritten work: *Les six livres de l'arithmetique by Diophante d'Alexandrie augmentez & reduits a la specieuse* [Ozanam, 1674?], which was never published and which, after being missing for 200 years, was found by Jean Cassinet (1928-1999) [Cassinet, 1986]. It has been studied in detail in [Gómez et al., 2021].

[16]Reference to the former Chancellor of France, Henry François D'Aguesseau.

what some French mathematicians had done or collected from Diophantus' work, referring to Ozanam's text as a great work at the level of Bachet's *Diophantus* published in 1621 ([Bachet, 1621]) and considered the standard Diophantus for decades, [17] or that of Father De Billy,[18] referring to the latter as a knowledgeable mathematician of the *Arithmetic* of Diophantus. Montucla wrote about Ozanam:

> M. Ozanam was also dedicating himself to this work [the Arithmetic of Diophantus]; and in Father De Billy's judgment, he had made an extraordinary effort. He had written a treatise on the analysis of Diophantus, which exists only in manuscript, and which M. Daguesseau possessed in 1717, according to what appears in the History of the Academy of Sciences in the eulogy [the *Eloge* by Fontenelle] of this author. This work would have contributed more to his reputation, not among the ordinary mathematicians, but among the skilled ones, than most of his other works.[19]

---

[17]The *Diophantus* published by Bachet in a bilingual edition in Greek-Latin, was considered from the beginning a great work. It became the reference book for anyone who, at the time, wanted to study the *Arithmetic* of Diophantus and, of course, for Ozanam as well. Bachet's text is written in two columns, on the right is the original Greek version and on the left is the Latin translation. Bachet's treatment was very rigorous and also innovative, and he did not limit himself to a mere translation of the Greek text; he made a reflection on the problems of Diophantus and obtained general solutions adding numerous comments and new problems from those of Diophantus himself. But Bachet did not use the new algebraic language at all.

[18]Jacques De Billy, a French Jesuit and mathematician, considered an expert connoisseur of the work of Diophantus, had J. Ozanam as a disciple and was related to Bachet. He also maintained a frequent correspondence with Fermat. Part of the comments and mathematical observations from Billy to Fermat, were collected under the title of "Doctrinae analyticae inventum novum" in the edition of the *Arithmetic* of Diophantus published by Fermat's son. Niceron [Niceron, 1739], Heath [Heath, 1910] and Meskens [Meskens, 2010], offer more details on De Billy's biography and works, as well as on his relationships with Ozanam, Bachet or Fermat.

[19]"M. Ozanam se jettoit vers le même tems dans cette carriere; et au jugement du P. De Billi, il y prenoit un effor extraordinaire. Il avoit écrit un Traité de l'analyse de Diophante, qui n'existe qu'en manuscrit, et que possédoit M. Daguesseau en 1717, suivant ce que nous apprend l'Historien de l'Académie des Sciences dans l'éloge de cet Auteur. Cet ouvrage eût contribué davantage à sa réputation, non auprès du vulgaire des Mathématiciens, mais auprès des habiles gens, que la plûpart de ceux qu'on a de lui", [Montucla, 1758, p. 321].

Montucla was referring to the *Eloge* of Fontenelle that we have seen before and, again, he emphasized the importance that Ozanam's *Diophantus* would have had at that time. In this case, he was relying on the opinion of another important mathematician, Father De Billy. Furthermore, he emphasized that this unpublished work would have increased recognition of him as an unusually skilled mathematician.

We have already seen that Leibniz praised Ozanam's *Treatise on Algebra*. However, both, Leibniz and Ozanam had long-standing relationship and so were their complementary remarks. Dated February 26, 1672 (February 16 in today's Gregorian calendar), Leibniz addressed a letter to Henry Oldenburg (1619-1677), secretary of the Royal Society of London, in which he said that Ozanam was living in Paris and was a very prepared young man who was applying algebraic methods to solve Diophantus' problems and was managing to solve problems whose solutions were not known:

> Mr Ozanam is a young man in Paris, highly skilled in algebra, who is to give us something of that sort (the same as was developed by Diophantus), having found a way of solving problems that could not be solved by the Diophantine or any algebraic method known hitherto.[20]

On October 16, 1674 (October 6 in the current Gregorian calendar), Leibniz addressed another letter to Oldenburg in which he spoke explicitly of the manuscript and, among other things, said that he had been able to see a manuscript by Ozanam on the *Arithmetic* of Diophantus. Leibniz gave a brief idea that Ozanam had written the manuscript in the symbolic language of the new algebra and that Father De Billy praised Ozanam's work. Leibniz affirmed that Ozanam had added numerous questions that had not been collected neither by Diophantus nor by Bachet; and given the novel, comprehensive, and rigorous character of Bachet's *Diophantus*, already commented, this statement is a clear example of the quality that Leibniz observed in this work by Ozanam.

---

[20]Translation by Hall and Hall [Hall, 1977] from the original text: "Parisiis est dominus Osannam juvenis in Algebra versatissimus, qui nobis aliquid in eo genere Idem Diophantum promotum dabit, reperta ratione solvendi problemata, quae neque ex Diophanto, neque ex cognita hactenus Algebra poterant solvi".

Although Leibniz claimed that the text would soon be published, it was never actually edited:

> Jacques Ozanam, about whom I at some time spoke to you, and whom Père Billy has mentioned with praise in his writings, showed me his work on Diophantos, soon to be committed to the press, reduced into symbolism. He has added throughout questions omitted by Diophantus and Bachet and he has added a seventh book filled with supplementary questions ... [21]

In his thoughts, Leibniz stated that the algebraic language allowed Ozanam to generalize the problems originally proposed by Diophantus of Alexandria, without being subject to the purely numerical; he even suggests to the Royal Society to publish a symbolic version of Diophantos:

> I think that the Diophantos of the same Ozanam will be worth reading; as he gives the work in such a way that he gets rid of all the lemmas sought from the nature of numbers, and shows they can always be found out by his analytical method. These are the matters which I judged worth writing, indeed, because I understand that a symbolic [version of] Diophantos is also to be, or perhaps is, published with you. [22]

Years later, in 1686, Leibniz referred again to the *Diophantus* of

---

[21] Translation by Hall and Hall [Hall, 1977] from the original text collected by Wallis [Wallis, 1699]: "Jacobus Osanna de quo TIBI aliquando locutus sum, et cuius P. Billy in scriptis suis cum elogio meminit, monstravit mihi super Diophantum suum mox praelo committendum; ad symbola revocatum. Adjicit passim Quaestiones a Diophanto et Bacheto praetermissas; sed et librum septimum addet refertum quaestionibus Paralipomenis". The seventh book with complementary questions which Leibniz referred to is not found in the manuscript, [Ozanam, 1674?], which we have studied in [Gómez et al., 2021].

[22] Translation by Hall and Hall [Hall, 1977] from the original text collected by Wallis [Wallis, 1699]: "Diophantum ipsius Osannae puto fore lectu dignum; dat enim operam, ut lemmata omnia ex numerorum natura petita, expungat, et ut semper ostendat ipsum inveniendi modum analyticum. Sed haec quidem vel ideo scriptu digna putavi, quia Diophantum symbolicum apud Vos quoque edi editumve esse intelligo".

Ozanam, which evidently remained unpublished, praising this work[23] in a letter addressed to Simon Foucher (1644-1696).[24]

> [...]Since I left Paris, from Monsieur Ozanam I have only seen his book on Practical Geometry, his Trigonometry and his new Gnomonic. I wait for what he will give us on Diophantus. This is where he could give something good.[25]

Let us remark that this letter goes on to expose a complaint from Lebniz to Ozanam; in fact, we can read:

> I found that he has not been very considerate of me; because he has inserted in his Géométrie my squaring of the circle (namely, being 1 the diameter, the [area of the] circle is $1+\frac{1}{3}+\frac{1}{5}+\frac{1}{7}+\frac{1}{9}$ etc. [sic])[26] with my demonstration, without naming me, and speaking with an air as if this demonstration were his.[27]

---

[23]More detailed information about the *Diophantus* of Ozanam can be found at [Gómez et al., 2021].

[24]Simon Foucher, canon of the cathedral at Dijon, was an important philosopher of the 17th century, who maintained a profuse epistolary relationship with Leibniz, see [Foucher, 1854].

[25][Foucher, 1854], p. 48 (see also AAII, 2, p. 87, in the Leibniz's Akademie official edition, https://www.uni-muenster.de/Leibniz/DatenII2/II2_B.pdf): "Depuis que j'ay quitté Paris, je n'ay vû de M. Osannam que son livre de la Géométrie pratique, sa Trigonométrie et sa nouvelle Gnomonique. J'attends ce qu'ils nous donnera sur Diophante. C'est là où il pourrait donner quelque chose de bon".

[26]There is a mistake in Foucher's translation, [Foucher, 1854, p. 48]; of course, Leibniz refers to the arithmetical quadrature formula

$$\frac{\pi}{4} = 1 - \frac{1}{3} + \frac{1}{5} - \frac{1}{7} + \frac{1}{9} - \frac{1}{11} \cdots$$

for the area of a circle with diameter 1.

[27]"J'ay trouvé qu'il n'a pas trop bien usé à mon égard; car il a inséré dans la Géométrie ma quadrature du circle (scavoir que diamètre estant 1, le cercle est

$$1 + \frac{1}{3} + \frac{1}{5} + \frac{1}{7} + \frac{1}{9} \text{ etc. [sic])}$$

avec ma démonstration sans me nommer et parlant d'un air comme si cette démonstration estoit de luy", [Foucher, 1854, p. 48].

Indeed, two years earlier, in 1684, Ozanam published in his *Géométrie Pratique* a proof for the arithmetical quadrature of the circle,[28] and he obtained that

$$4aa - \frac{4}{3}aa + \frac{4}{5}aa - \frac{4}{7}aa + \frac{4}{9}aa + \cdots$$

is the area of a circle whose diameter has length $2a$.[29]

Time passes and we have no news in the correspondence about this dispute until the end of 1691, when Foucher writes to Leibniz in a letter dated December 30, 1691,[30] to communicate a words of Ozanam in which the French mathematician assures that, of course, Leibniz gave him the *overture* to the squaring of the circle; but, faced with the problem, he spent so much time in this labour, until the final discovery, that he claims also the right for such a discovery:

> Mr. Ozanam says it is true that you have gave him the overture of his squaring of the circle; but he complains that, having been too slow to discover what you know of it, you have given him reason to do his meditations on it, and to find out what he has found of it, so that he claims to be entitled, as well as you, to this discovery; but with all that, I would have wished he had named you.[31]

Inside this correspondence, subsequently Leibniz writes, in 1692, to

---

[28][Ozanam, 1684, pp. 192-200].

[29]For a proof of the construction of the formula by Leibniz, as well as for having a general scope about the problem of squaring the circle in the seventeenth century, see [Crippa, 2014], [Crippa, 2019], or the Leibniz's original edited by Knobloch in 1993, [Leibniz, 1993]. We must take into account that, according to Crippa, Leibniz's work *De quadratura arithmetica circuli ellipseos et hyperbolae cujus corollarium est trigonometria sine tabulis* was composed and ultimated during the Leibniz's stay at Paris from 1672 to 1676, but published only in Knobloch's edition of 1993.

[30][Foucher, 1854, p. 85].

[31]"M. Osannam dit qu'il est vray que vous luy avez donné l'ouverture de sa quadrature du cercle; mais il se plaint de ce qu'ayant esté trop lent à découvrir ce que vous en scavez, vous lui avez donné lieu de faire là-dessus ses méditations, et d'en trouver ce qu'il en a trouvé, de sorte qu'il prétend avoir droit, aussi bien que vous, à cette découverte; mais avec tout cela, je voudrois qu'il vous eust nommé", [Foucher, 1854, p. 85].

Foucher[32] in these terms:

> Mr. Ozanam will not deny that I gave him the first views of the squaring of a circle, of which we have spoken, him & me; & I would have communicated my demonstration to him, if he had asked me.[33]

Nevertheless, the letter continues with a positive assessment of Ozanam's *Dictionnaire* and, at the same time, we know that Leibniz showed to Ozanam the use of equations for geometric constructions:

> He will also admit that I am the first who showed him the use of local equations for the constructions, with which he was delighted. He has made a very good use of them, as I see by his Dictionary.[34].

Additionally, Leibniz thinks that Ozanam could be able to do certain analytic tables because "he is one of the men in the world who has the most ease & practice for the ordinary calculation of the specious, I had thought that something as useful as this could be done by his means".[35] To end the passage of this discussion about the authorship of the arithmetic quadrature of the circle, we mention Ozanam's respect for the figure of Leibniz through the words that Foucher addresses to Leibniz himself, in a letter dated in August 1692, in which he says that Ozanam "kiss your hands" and, at the same time, proposes to Leibniz another problem.[36]

---

[32]See [Foucher, 1854, pp. 88-92]; this letter was sent with modifications to *Journal des Savants*, June 2, 1692, [Journal, 1692].

[33]"M. Ozanam ne disconviendra pas que je ne luy aye donné les premières vuës de la quadrature du cercle, dont nous avons parlé lui, et moi, et je lui en aurois communiqué ma démonstration s'il me l'avoit demandée", [Foucher, 1854, p. 90].

[34]"Il avouëra aussi que je suis le premier qui lui ai montré l'usage des equations locales pour les constructions, dont il fut ravi. Il en a fait un fort bel usage, comme je voi par son Dictionnaire", [Foucher, 1854, p. 90][Journal, 1692].

[35]"Mr Ozanam est un des hommes du monde qui a le plus de facilité & de pratique pour le calcul ordinaire de la specieuse, j'avois pensé qu'une chose aussi utile que celle-là se pourroit faire par son moyen", [Foucher, 1854, pp. 90-91].

[36]"M. Osannam vous baisse les mains; il m'a donné un problème pour vous, je le mets icy tel qu'il me l'a donné" [Foucher, 1854, p. 96].

To finish this section, we can conclude that Ozanam was an outstanding mathematician who knew the mathematical works published in his time and used them in practice. Furthermore, his work was widely known and valued in light of the numerous reissues and praise it received.

2.3   On his unpublished manuscript

In addition to the conjectures proposed in [Gómez et al., 2021], we have found another plausible reason for the missed publication of Ozanam's manupscript. It concerns to the figure of Jean-Baptiste Colbert (1619-1683), First Minister of State of France from 1661 until his death, under the auspices of Louis XIV.[37]

Colbert, as we can imagine, was a very influential and powerful person at the court of Louis XIV. It is not surprising, thus, the large number of dedications that Colbert received from different authors, for instance: the book *Diophanti Alexandrini Arithmeticorum libri sex, et De numeris multangulis liber unus. Cum commentariis C.G. Bacheti V.C. & observationibus D.P. de Fermat senatoris Tolosani.* edited under the care of Fermat's son Samuel in 1670, in which, in addition to Bachet's translation of Diophantus' Arithmetic he included his father's notes; the same Ozanam dedicated to Colbert his book *Traité de Gnomonique*, [Ozanam, 1673]; or De la Hire and his *Nouvelle Methode en Geometrie pour les sections coniques & cylindriques*, Thomas Moette, Paris, 1674. Colbert founded the *Académie Royale des Sciences* in 1666, and from the first time he tried to attract the best scientists of the epoch, as for instance Cassini, Huygens, or Leibniz.[38]

In this context, the support of an influential maecenas was decisive for prospering both socially and professionally. And Ozanam, maybe due to the precedent of the book *Diophanti Alexandrini Arithmeticorum*

---

[37]For a biographic account of Colbert, see [Clément, 1846].

[38]For a general scope of the *Académie*, apart from the Clément's book signaled in the precedent footnote and which places information about the creation of the *Académie*, you can consult for instance, *Index biographique de l'Académie des Sciences (1666-1978)*, Paris: Gauthier-Villars, 1979; for the expenses necessary to trap the talent of reputed scientists, consult [Stroup, 1987]; also, consult [Stroup, 1996], where, among others, it depicts some of Huygens's achievements attained during the period in which he was one of the leading members of the *Académie des Sciences*. In the case of Leibniz, consult [Salomon-Bayet, 1978].

libri sex, hoped Colbert's support would be significant in bringing his manuscript to light. Unfortunately, Colbert's death in 1683 shattered such hope. According to the preface of Ozanam's *La Géométrie pratique* from 1684, the mathematician could have received Colbert's approval to publish his manuscript, but, unfortunately, Colbert's death perhaps truncated the project (see Figure 3):

> The unforeseen death of Monsieur Colbert having broken his desire which he had to do Diophantus' Arithmetic printed, which I increased & reduced to Specious, with three other Treatises, whose project has been published, makes me work hard again on the six Books of this Author, which I will increase to at least double what they were previously, so as not deprive the Public of the new thoughts which continually come to me on this matter. I stopped working on it during the time I spent composing this Book, and another which is now in the Press, which deals with Rectilinear & Spherical Trigonometry, & the construction of the Sine & Logarithms Tables, which could not dispense with at the request of the Librarian.[39]

Some years later, in 1688, Ozanam took advantage of the occasion[40] of citing and promoting his *Diophantus* in [Ozanam, 1688(b), pp. 101-

---

[39]"La mort imprevue de Monsieur Colbert ayant rompu le dessein qu'il avoit de faire imprimer l'Arithmetique de Diophante, que j'ay augmentée & reduite à la Specieuse, avec trois autres Traitez, dont le projet a été publié, me fait travailler tout de nouveau sur les six Livres de cet Auteur, que j'augmenteray pour le moins du double de ce qu'ils étoient auparavant, pour ne pas priver le Public des nouvelles pensées qui me viennent continuellement sur cette matiere. l'ay discontinué d'y travailler pendant le temps que j'εy employé à composer ce Livre, & un autre qui est à present sous la Presse, qui traite de la Trigonometrie Rectiligne & Spherique, & de la construction des Tables de Sinus & des Logarithmes, dont je n'ay pas pu me dispenser à la priere que m'en a faite le Libraire". Au Lecteur, [Ozanam, 1684, s.n.]. See Figure 3.

[40]"Quoyque ce ne soit pas icy le lieu de mettre cette Question, qui manque à 22.2. Dioph. outre que j'en ay déja enseigné la solution à plusieurs personnes d'esprit, qui ont été mes disciples dépuis plusieurs années, sçavoir en l'année 1680", [Ozanam, 1688(b), p. 101] // "Although this is not the place to put this Question, which lacks to 22.2 Diophantus [Question XXII, Book 2], besides that I have already taught the solutions to several intelligent people who have been my disciples for several years, namely in the year 1680".

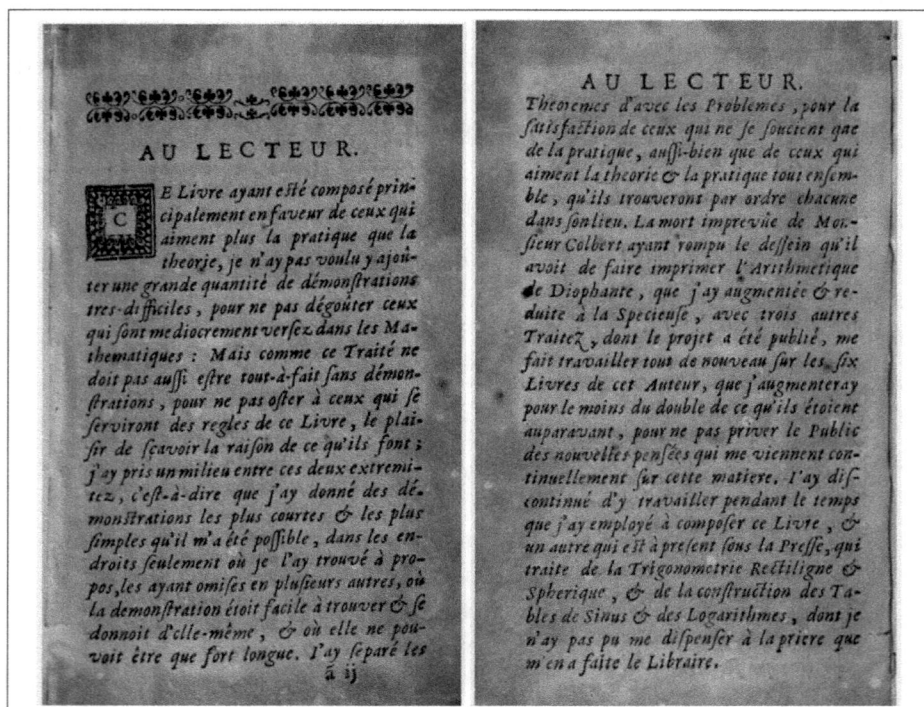

Figure 3: Preface of [Ozanam, 1684] with the news of Ozanam's *Diophantus*. (see: http://digital.onb.ac.at/OnbViewer/viewer.faces?doc=ABO_%2BZ179882805).

115]. Furthermore, Ozanam mentioned again the difficulties he had until then to publish his *Diophantus* and insisted that he did not lose hope that his book would be published,[41] and justified the inclusion of the question -to wit, to find two numbers so that the differences between each number and the square of the other one give square numbers- based on the facts of enlarging the book ("pour rendre ce livre ([Ozanam, 1688(b)]) un peu plus ample") and of demonstrating to the public his ability acquired behind a long study of the Algebra ("pour faire partie

---

[41] See [Ozanam, 1688(b), pp. 102,114-115]: "[...] par la difficulté que j'ay trouvée jusques à present à faire imprimer ma Diophante [...] en attendant que mon Diophante paroisse... Il y a de belles reflexions à faire là dessus, que vous trouverez dans nôtre Diophante, lors qu'il aura le bon-heur de parôitre...".

au Public des lumieres que j'ay acquises par une longue etude dans l'Algebre").

With this note, we would like to give another self-quote by Ozanam from his own work, not listed in [Gómez et al., 2021]. As a byproduct, we highlight the important pedagogical labour of Ozanam in the mathematical formation of young spirits in his time. Indeed, in the preface of [Ozanam, 1688(b)] we have the opportunity to see the names of some of his disciples.[42] He refers to Lucas, Dumourron and Sedilo. With regard to this last disciple,[43] whose name is actually recognized as Sédileau, not much information is known: in addition to pointing out his membership in the *Académie Royale des Sciences* since 1682 (realize that, in turn, Ozanam was a member of the Académie from 1702 to his death), we must mention that he worked mainly with De la Hire and Cassini, dealing with astronomical questions as well as some of the natural sciences related to the description of certain insects or the evolution of fluvial flows.[44] Unfortunately, Sédileau died in April 1693 and, according to Ozanam,[45] he was his student in 1680, then we can infer that he passed away very young.

3 Ozanam's mathematics and algebra

To go into some detail about Ozanam's way of working, we will take a look at some of the issues he addresses in his *Diophantus*. We will

---

[42]...qui ont été mes disciples dépuis plusieurs années, sçavoir en l'année 1680. à Messieurs Lucas & Dumourron, & particulierement à Monsieur Sedilo, lequel depuis ce tems-là a été receu par son merite dans l'Academie Royale des Sciences. [Ozanam, 1688(b), pp. 101-102].

[43]We have not found any information about the two other disciples.

[44]The reader interested in the Sédileau's work can consult *Table alphabetique des matieres contenus dans l'Histoire & les Memoires de l'Académie Royale des Sciences. Tome Premier*, edited under the care of L. Godin, Paris: Compagnie des Libraires, 1730, where a list of his works is provided jointly with the corresponding volumes of the *Memoires de l'Académie Royale des Sciences* in which these works are reproduced, in particular, the tenth volume published at Paris in 1730 also by the Compagnie des Libraires.

[45]"[...] qui ont été mes disciples dépuis plusieurs années, sçavoir en l'année 1680. à Messieurs Lucas & Dumourron, & particulierement à Monsieur Sedilo", [Ozanam, 1688(b), p. 101] // "[...] who have been my disciples for several years, namely in the year 1680. to Mr. Lucas & Dumourron, & particularly to Mr. Sedilo".

try to show his mastery of language and algebraic techniques, as well as the versatility of his mathematical thinking that allows him to tackle problems of different kinds by relating them and finding solutions in different areas: numerical, arithmetic, algebraic or geometric. In order to demonstrate the power of algebraic language and Ozanam's thought, we will show how these problems were addressed in Bachet's extraordinary *Diophantus* [Bachet, 1621] which does not use the new algebraic language, but is instead written in a rhetorical and cosist language.

3.1 Ozanam's notation

The algebraic notation used by Ozanam basically follows the one introduced by Descartes [Descartes, 1637] and, although with some differences, it was very similar to the present one. It should be borne in mind that the new algebraic language started at the beginning of the 17th century. However, the process of evolution, until the use and meaning of the symbols became common, was relatively short considering the profound change it brought about in the evolution of mathematics.

Ozanam, again following what was done by Descartes, reserved the letter l to designate the unit that he would use to equal "the size" of the quantities. This "size" had to do with the geometric dimension, that is, Viète's law of homogeneous: an equality could not be established between two quantities of different dimensions. This is how Ozanam himself explained it:

> This is done to preserve the law of homogeneous, that is, in order not to deviate from the rules of geometry, which teaches us that there is no ratio between a line [dimension 1, for example x] and a plane [dimension 2, for example x^2], or between a plane and a solid [dimension 3, for example x^3], because these quantities are heterogeneous, that is, of different kinds; for in this way we can solve any arithmetic problem by geometry, as you will see in the first two books.[46]

---

[46]"Cela se pratique pour conserver la loy des Homogenes, c'est à dire pour ne point s'éloigner des regles de la Geometrie, qui nous aprend qu'il n'y a aucune raison entre une ligne & un plan, ny entre un plan & un solide, & c. parce que ces grandeurs sont heterogenes, c'est à dire de different genre; car ainsy on peut resoudre tout probleme

Ozanam also explained that he would use the word "equation" instead of "equality" because an equation was the comparison of two different quantities to make them equal, whereas an "equality" was a comparison between two quantities that are already identical.[47]

A comparison can be observed between the notation used by Viète, Descartes and Ozanam in Table 1 ([Massa-Esteve, 2001] and [Gómez et al., 2021]).

Signs	Viète (1590s)	Descartes (1637)	Ozanam (1700)
Equality	æqualis	$\propto$	$\sim$, ∞
Greater than	*Maior est*	*Plus grande*	$\oplus$
Less than	*Minus est*	*Plus petite*	$\ominus$
Product of a and b	A in B	ab	ab
Addition	*plus*	$+$	$+$
Subtraction	*minus*	$-$	$-$
Ratio	*ad*	à	$::$
Square root	$VQ.$	$\sqrt{}$	$\sqrt{}$
Cubic root	$VC.$	$\sqrt{c}$	$\sqrt[3]{}$
Squares	A *quadratus*, A *quad*	a^2, aa	a^2, aa
Cubes	A *cubus*, A *cub*	a^3	a^3
Absolute value			$\cdots$

Table 1: Comparative notation table among Viète, Descartes and Ozanam.

Other symbols used by Ozanam:

$\sim$, ∞	Denotes equality, the current sign $=$.
$\oplus$	Denotes "greater than"; the actual $>$.
$\ominus$	Denotes "less than"; the actual $<$.
$\cdots$	Between two quantities, it is similar to the current absolute value.

Moreover, he used the symbol $::$ to express the ratio between two

---

d'Arithmetique par Geometrie, comme vous verrez dans les deux premiers livres", [Ozanam, 1674?, p. 5].

[47]For more information on the concept of equation, see Cajori [Cajori, 1993].

quantities, that is, $x, y :: a, b$ meant that x and y are in the same ratio as a and b, that is $\frac{x}{y} = \frac{a}{b}$ in the current notation. He denoted the square powers as xx in the manuscript, but also as x^2 in other works, like Descartes.

3.2   Ozanam's Question XVIII of Book VI

Let us present the development of Question XVIII in Ozanam's manuscript [Ozanam, 1674?, Livre VI, pp. 409-410]. The problem is concerned with the searching of right triangles ABC whose sides AB, AC, BC and the bisector AD from an acute angle can be expressed as rational numbers.

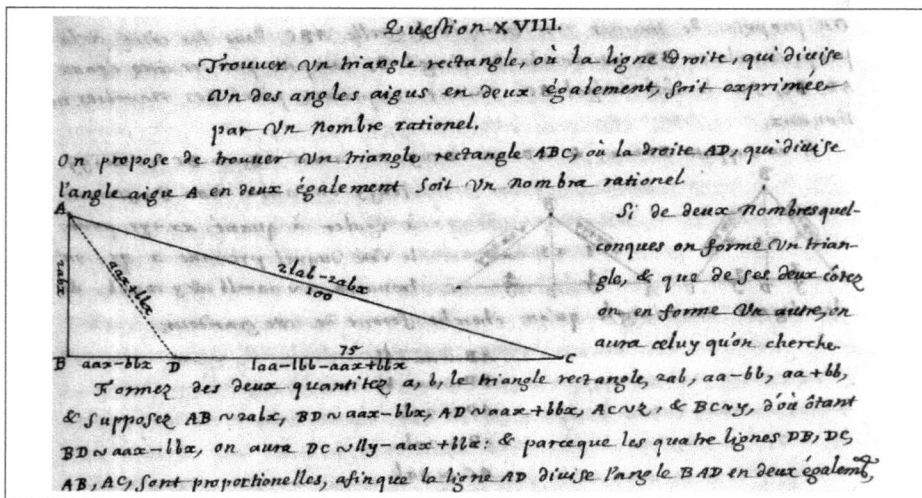

Figure 4: Part of the Question XVIII, Diophantus' Book VI, in the Ozanam's manuscript.

Ozanam's strategy consists in constructing at first place a right triangle ABD and after it in expanding this minor triangle to the desired triangle ABC.

He takes two arbitrary quantities a, b and, according to well-known Pythagorean triples, considers the right triangle having sides $2ab$, $a^2 - b^2$, $a^2 + b^2$; next, he takes (see Figure 4) the values

$$AB = 2abx, BD = a^2x - b^2x, AD = a^2x + b^2x$$

and denotes $AC = z$, $BC = y$. Consequently, we have[48] $DC = y - a^2x + b^2x$.

He continues by using the well-known fact related with the bisector, to wit, the segments DB, DC, AB, AC are proportional, that is $\frac{DB}{DC} = \frac{AB}{AC}$, therefore[49] $\frac{a^2x - b^2x}{y - a^2x + b^2x} = \frac{2abx}{z}$. From here, $z = \frac{2aby}{a^2 - b^2} - 2abx$.

Now, Ozanam chooses $y = a^2 - b^2$ in order to simplify the numbers and to get integer numbers, since in this case we have $z = 2ab - 2abx$. In this way, Ozanam has obtained the value of each line in the triangle of Figure 4, AB, AD, AC, BD, DC. It only remains to impose that the triangle ABC is, in fact, a right triangle. To this end, Ozanam constitutes an equation in the indeterminate x by equating $AC^2 = AB^2 + BC^2$ or $(2ab - 2abx)^2 = (2abx)^2 - (a^2 - b^2)^2$. It is easily seen that $x = \frac{6a^2b^2 - a^4 - b^4}{8a^2b^2}$.

Finally, Ozanam gives the general solution ("on aura en entiers")[50]

$$AB = 12a^3b^3 - 2a^5b - 2ab^5,$$
$$AC = 4a^3b^3 + 2a^5b + 2ab^5,$$
$$AD = 5a^4b^2 + 5a^2b^4 - a^6 - b^6,$$
$$BC = 8a^4b^2 - 8a^2b^4,$$
$$BD = 7a^4b^2 - 7a^2b^4 - a^6 + b^6,$$
$$CD = a^4b^2 - a^2b^4 + a^6 - b^6.$$

In particular, if $a = 2$, $b = 1$, he obtains the values appearing in Figure 4 (multiplied by $8a^2b^2$). For instance, $CD = 16 - 4 + 64 - 1 = 75$.

---

[48]Notice that Ozanam employs the notation $DC \sim \ell\ell y - aax + bbx$, here we can find Descartes' influence concerning the writing of squares and we observe that the law of homogeneous is preserved by introducing the quantity ℓ as many times as necessary.

[49]To indicate the proportion, Ozanam writes $aax - bbx, \ell\ell y - aax + bbx :: 2abx, \ell\ell z$.

[50]Ozanam uses the expression "On aura en entiers"/"We will have in integers" to denote that each line must be divided by $8a^2b^2$; for instance, $BC = 8a^4bb - 8aab^4$ is meant the fraction $BC = \frac{8a^4bb - 8aab^4}{8a^2b^2}$.

3.3 Ten problems added to the Question XVIII of Book VI in Ozanam's manuscript

Once Ozanam finished the solution of Question XVIII of Book VI, he affirmed:

> On the occasion of this Question, here we will add the following ones, the first six questions belonging to Bachet, who has solved them partially, except the first three problems, because in the others he did not make a rational perpendicular.[51]

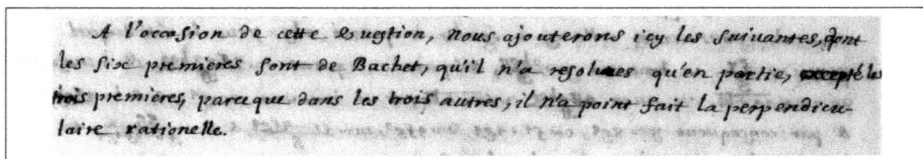

Figure 5: Ozanam adds ten problems to Question XVIII.

The ten added problems are concerned with the searching of triangles with rational numbers for the sides of the triangle and its perpendicular, except for Problems VIII and IX dealing with sides and diagonals of quadrilaterals.

For example the Problem VII is:

> Find a triangle, whose three sides & one perpendicular are four rational numbers in a continual arithmetic proportion.[52]

Or the Problem VIII is:

---

[51]"A l'occasion de cette Question, nous ajouterons icy les suivantes, dont les six premieres sont de Bachet, qu'il n'a resolues qu'en partie, excepté les trois premieres, parce que dans les trois autres, il n'a point fait la perpendiculaire rationelle", [Ozanam, 1674?, Livre VI, p. 410].

[52]"Trouver un triangle, dont les trois côtez & une perpendiculaire soient quatre nombres rationaux dans une continuelle proportion arithmetique", [Ozanam, 1674?, p. 420].

Find a quadrilateral in a circle, of which the four sides, and the two diagonals, are expressed by rational numbers.[53]

It is interesting to mention that Ozanam followed the order of Bachet's book when writing the statements of different parts of Diophantus' work. His merit is to prove them in a modern algebraic language, similar to our current procedures. This improvement allowed him to obtain also new results as the new Problems VII-X show. Moreover, it is relevant to notice that Ozanam mixed both algebraic and geometric points of view, as new Problem X will reveal.

3.3.1 Problem I of Question XVIII of Book VI: Bachet and Ozanam procedures

Let us explain briefly how Ozanam solved the new Problem I added to Question XVIII (see Figure 6 to follow the development):

Find an isosceles triangle, whose three sides and the perpendicular which falls from one of the angles on its opposite side, are expressed by rational numbers.[54]

On the triangle isosceles ABC, with perpendicular BD, he appoints $BD = x$, $AD = CD = y$, and therefore $AB = BC = \sqrt{x^2 + y^2}$. Since the sides AC, BC and the perpendicular BD must be rational, he takes $x^2 + y^2 = z^2$, to be more precise $x^2 + y^2 = (x - \frac{a}{b}y)^2$. By expanding the square and simplifying, we deduce $y(a^2 - b^2) = 2abx$. Ozanam then establishes, merely by comparing, $x = a^2 - b^2$, $y = 2ab$. Therefore, he affirms that the searched lines can be given as $AB = a^2 + b^2$ (notice that, in fact, it is necessary to think that $x^2 + y^2 = (\frac{a}{b}y - x)^2$ because $\frac{a}{b}y = 2a^2 > a^2 - b^2 = x$); $BD = a^2 - b^2$; $AC = 4ab$ since $AC = 2AD = 2CD = 2y = 4ab$.

In particular, if we put $x = 4$ and $y = 3$ (see Figure 6), we obtain the example examined by Bachet (see Figure 7).

---

[53]"Trouver une quadrilatere dans un cercle, dont les quatre côtez, & et les deux diagonales, soient exprimez par des nombres rationaux", [Ozanam, 1674?, Livre VI, p. 421].

[54]"Trouver un triangle isoscelle, dont les trois côtez, & la perpendiculaire, qui tombe d'un des angles sur son côté opposé, soient exprimez par des nombres rationaux", [Ozanam, 1674?, Livre VI, p. 410].

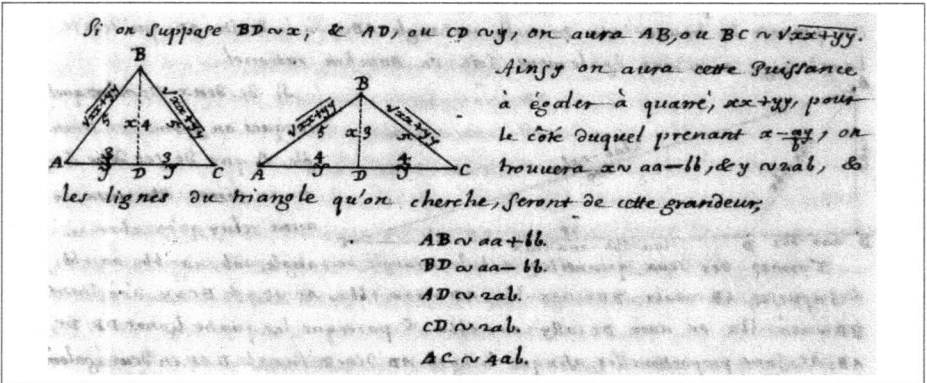

Figure 6: Problem I in Question XVIII by Ozanam.

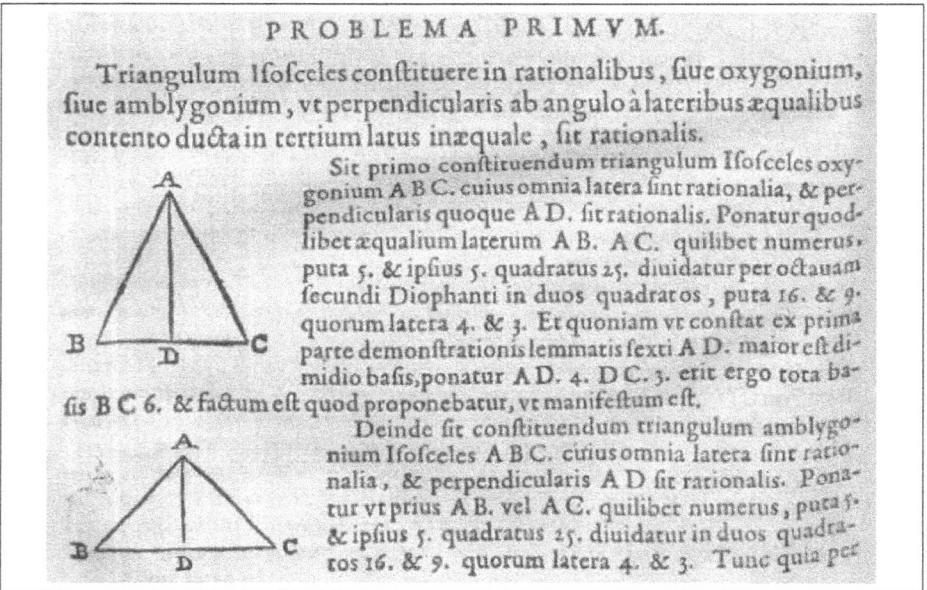

Figure 7: Problem I in Question XVIII by Bachet. [Bachet, 1621, p. 414], image from: https://www.e-rara.ch/zut/content/zoom/2568771.

Let us comment on Bachet's procedure to check that, indeed, we obtain the numbers prognosticated by Ozanam. Notice, firstly, that Bachet distinguishes between "Triangulum oxygonium" -the angle between the two equal sides is acute- and "Triangulum amblygonium" -now, the angle is obtuse. For the "oxygonium" case, in the triangle ABC Bachet assigns to the equal sides an arbitrary number, say $5 = AB = AC$. Following that, to its square (here, 25) he applies the solution of Problem VIII in Book II of Diophantus, so it is possible to decompose 25 as the sum of two square numbers (in this case, the decomposition is immediate, $25 = 9 + 16$), and consider the corresponding sides of these squares (in the example, 3 and 4, respectively). Being "oxygonium" the triangle, by suitably combining results and arguments from Euclid's *Elements*, Bachet knows that the perpendicular line is greater than half the base BC (he has proved it previously jointly with a series of lemmas that preceded Problem I). For this reason, he takes $AD = 4$ and $DC = BD = 3$ (remember that 3 and 4 are the sides of 9 and 16 in the decomposition of 25 as sum of squares). Finally, he solves the problem with $BC = 6$.

The reasoning for the "amblygonium" case is analogous, but now the perpendicular must be less than half the base BC (Bachet also proves it previously). In this way, taking again $5 = AB = AC$, now the isosceles triangle is determined by taking $AD = 3$ and $DC = BD = 4$, therefore $AD = 3$ and $BC = 8$.

We return to Ozanam's resolution of Problem I in Question XVIII. Once Ozanam has described the solution (recall, $AB = a^2 + b^2$, $BD = a^2 - b^2$, $AC = 2AD = 2CD = 4ab$), he discusses the possible nature of the isosceles triangle, either *oxygonium* or *amblygonium*.

In the first case, he remarks, but without proof, that it is necessary that $BD > AD$: "...ainsy on aura cette inégalité, $aa - bb \oplus 2ab$, dans laquelle on trouvera $a \oplus b + \sqrt{2bb}$, ..." (in this way we will have the inequality $a^2 - b^2 > 2ab$ from which we have $a > b + \sqrt{2}b$); notice that the root $b - \sqrt{2}b$ is not considered and realize that Ozanam is solving an inequality with positive numbers a, b. As an example, he considers $b = 1$, $a = 3$.

On the other hand, for the *amblygonium* case, now the necessary condition is $BD < AD$. Now, Ozanam writes (take into account that $\ominus$ is the notation used for our actual symbol $<$):

"...so we will have this inequality $aa - bb \ominus 2ab$, in which we will find $a \ominus b + \sqrt{2bb}$: & if we assume $b \sim 2$, & $a \sim 4$, the lines of the triangle we are looking for will be in lesser terms, of such magnitude that you see them marked in the second figure".[55]

We can see the difference between Ozanam and Bachet's approach to solve the problem. Bachet's cossist and rhetorical language, with a geometric point of view in the solutions, connected with Euclid's *Elements*, in contrast to the skill with which Ozanam uses the new algebraic language that allows him to have a more algebraic point of view, which even allows us to characterize algebraically the conditions that the variables a and b have to fulfill in order to obtain *oxygone* and *amblygone* triangles.

3.3.2 An improvement to Bachet's six problems accompanying Question XVIII

The **added problem V** to Question XVIII asks for a scalene triangle, either *oxygone* or *amblygone*, such that its sides, the perpendicular from one of the acute angles, and the bisector line be rational numbers.

Bachet's resolution, in a cossist and rhetorical language, is much more extensive than that carried out by Ozanam. Furthermore, Bachet needs to introduce some lemmas with partial results used to solve this Problem V. But, as we have seen in Figure 5, Ozanam affirms that Bachet does not obtain a rational perpendicular. Indeed, having been finished the resolution of Problem V, Bachet inserts a comment or annotation ("scholium"), namely: "we do not take care of whether the perpendicular AE is or not rational, because the problem only asks for the sides and the bisector line".[56]

Instead, Ozanam's resolution does find the sought rationals and, using algebraic language, he gets a much shorter and more elegant solution.

---

[55]"... ainsy on aura cette inégalité $aa - bb \ominus 2ab$, dans laquelle on trouvera $a \ominus b + \sqrt{2bb}$: & si on suppose $b \sim 2$, & $a \sim 4$, les lignes du triangle qu'on cherche, seront en moindres termes, de telle grandeur que vous les voyez marquées dans la seconde figure", [Ozanam, 1674?, Livre VI, p. 411].

[56]"Non curamus utrum perpendicularis AD sit rationalis, an non; cùm sufficiat latera trianguli unà cum linea AE facere rationalis", [Bachet, 1621, p. 420].

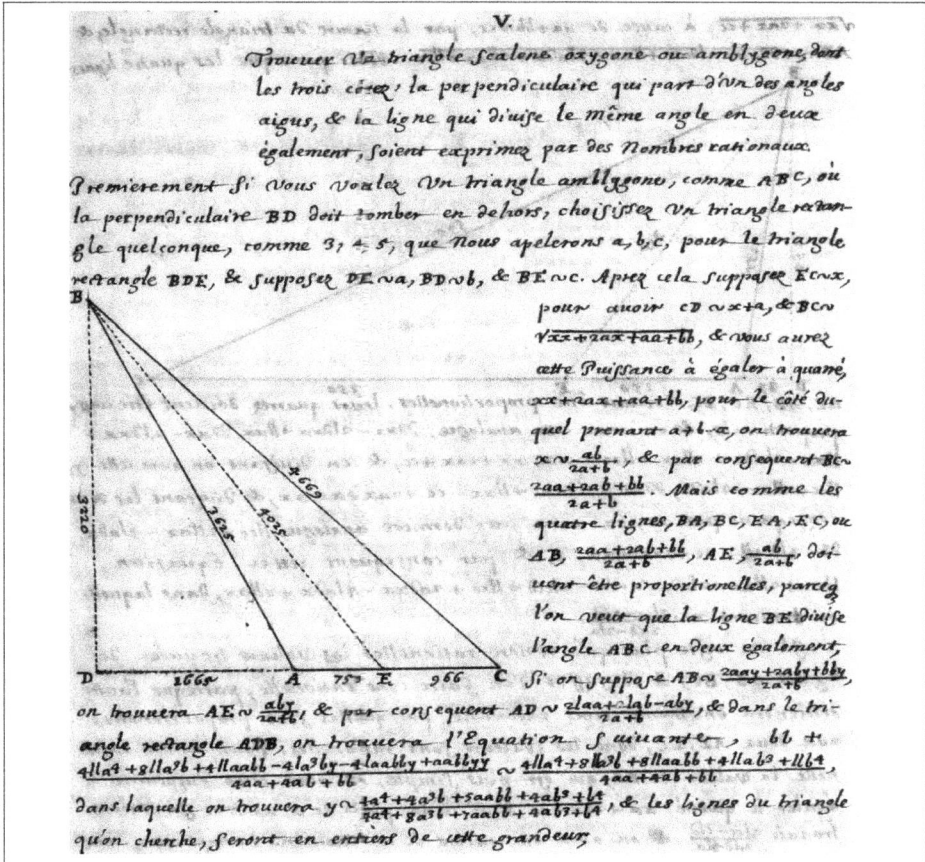

Figure 8: The case amblygone in Problem V of Question XVIII by Ozanam.

Ozanam divides the proof of this problem in two cases: *amblygone* triangle and *oxygone* triangle (see Figures 8 and 9, respectively).

For the *amblygone* case, at least let us mention that he traces the perpendicular outside the triangle and solves the question by two methods. The first of these methods is followed in the *oxygone* case, where Ozanam only sketches the main steps, because he applied the method in an exhaustive way when dealing with the *amblygone* case.

Thus, Ozanam stated that he completely solved Problem V using

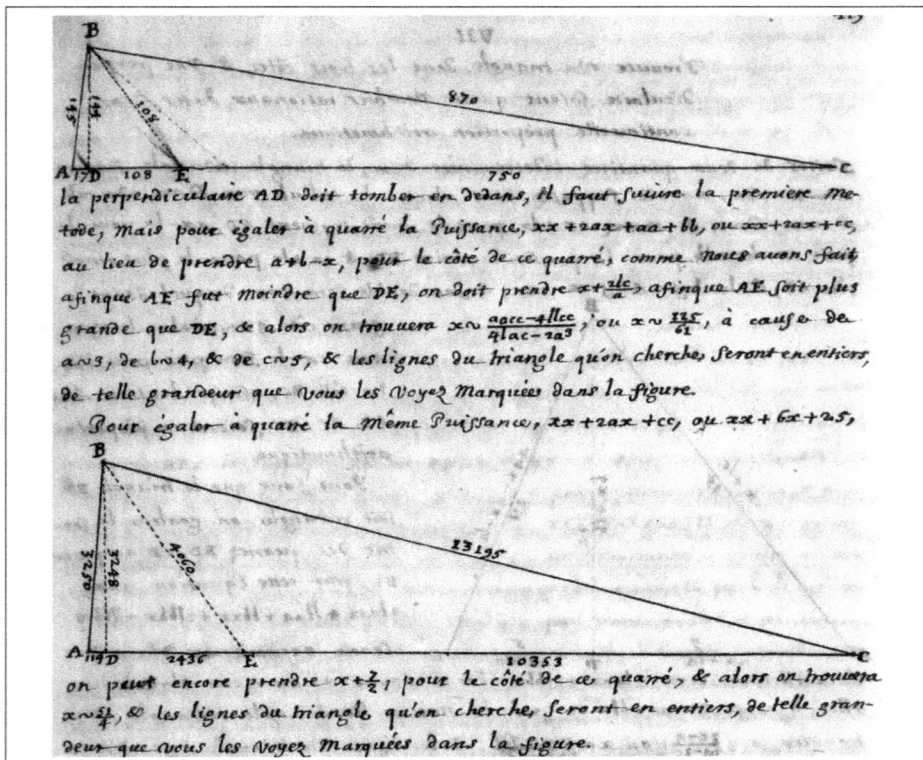

Figure 9: The case oxygone in Problem V of Question XVIII by Ozanam.

algebra, obtaining rational solutions as the formulation of the problem demanded, but by using algebra, he obtained general solutions from the conditions imposed by the Problem itself.

3.4   The relationship between algebraic solutions and the geometry

We are going to see here how Ozanam was able to obtain geometric consequences from the general solutions that he had obtained from the problems. Let us observe how, from problems with a geometric formulation such as triangles, he translates them into algebraic terms that allow him to extract general solutions, to later obtain geometric consequences that, we could say, are more complex than the starting triangles themselves. We are going to focus here on Problem X, added to Question XVIII.

This problem says (see Figure 10):

> Find two isosceles triangles with the same area & the same perimeter, whose sides & and perpendiculars are expressed by rational numbers.[57]

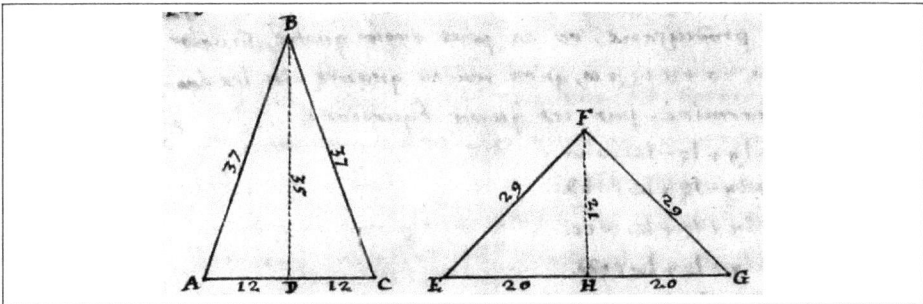

Figure 10: Drawing illustrating Problem X, [Ozanam, 1674?, Livre VI, p. 426].

Ozanam came up with two different solutions to the problem in the rigorous and elegant style that we have already seen, and he asked whether it is possible to solve the same question in the case of three or more isosceles triangles:

> This question was only proposed for two isosceles triangles because it is impossible to find three of them having the same perimeter, as it can be known by fixing the area and the perimeter, namely, by searching for an isosceles triangle whose area and perimeter be given, since we will find that the problem is solid, and the two loci serving up for the solution of the problem cannot be cut at anything other than two points which will provide two solutions, and consequently

---

[57]"Trouver deux triangles isosceles de même aire & de même contour, dont les côtez & et les perpendiculaires soient exprimez par des nombres rationaux", [Ozanam, 1674?, Livre VI, p. 425].

two isosceles triangles having the same area and the same perimeter.[58]

Ozanam proved this statement by using a geometric procedure. The new question reads as follows:

It is proposed to find an isosceles triangle ABC whose perimeter be equal to the given line AE and whose area be equal to the square of the given line AF[59] (see Figure 11).

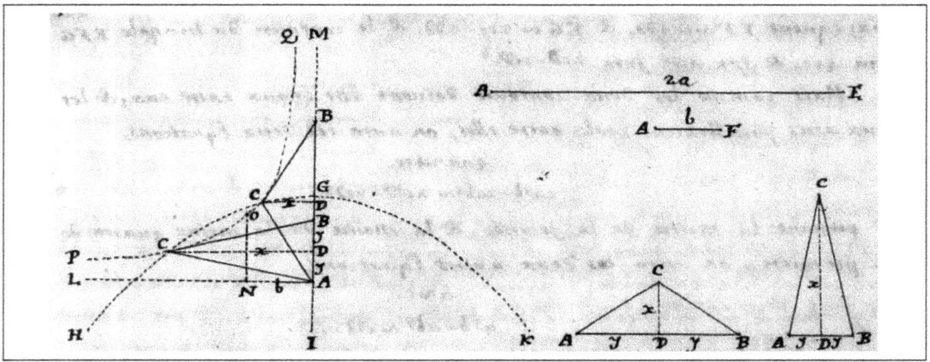

Figure 11: Geometric solution of the new Problem X in Question XVIII, [Ozanam, 1674?, Livre VI, p. 428].

Given an isosceles triangle ABC, Ozanam takes the segments AC and BC as being equal and, from the vertex C, draws the perpendicular CD on the base AB, therefore dividing AB in two equal parts from its point D. Next, he uses the notation $AE = 2a$, $AF = b$ for the known values, and $CD = x$, $AD = BD = y$ for the unknown quantities.

---

[58] "Cette question n'a été proposée que pour deux triangles isosceles, parce qu'il est impossible d'en trouver trois de même contour, comme l'on connoitra en determinant l'aire & le contour, savoir en cherchant un triangle isoscele, dont l'aire & le contour soient donnez car on trouvera que ce Probleme est solide, & que les deux Lieux, qui serviront pour la solution du Probleme, ne se peuvent couper qu'en deux poins, qui donneront deux solutions, & par consequent deux triangles isosceles de même aire & de même contour", [Ozanam, 1674?, Livre VI, p. 427].

[59] "Qu'il soit proposé de trouver un triangle isoscele ABC, dont le contour soit égal à la ligne donné AE, & dont l'aire soit égale au quarré de la ligne donnée AF", [Ozanam, 1674?, p. 427].

According to the requirement on the area imposed in the problem, it is necessary to have $xy = b^2$ (in Figure 11, the *hyperbola* corresponds to the dashed line joining the points P and Q); on the other hand, the condition on the perimeter yields to $2y + 2\sqrt{x^2 + y^2} = 2a$, since $AC = \sqrt{x^2 + y^2}$, thus, simplifying, $x^2 = a^2 - 2ay$, the equation of the locus which Ozanam identifies as the *parabola* (in Figure 11, the dashed line connecting points H and K) with *latus rectum* $2a$.[60] Thus, once the two loci (hyperbola and parabola) are presented, from their intersection, Ozanam will draw the corresponding geometric construction.[61]

Realize that Ozanam translates a geometric problem to algebraic language, by giving the equations of the loci, $xy = b^2$ and $x^2 = a^2 - 2ay$, and next he returns to the geometrical way by drawing the parabola and the hyperbola with the purpose of proving that, at most, only two values for x are possible, just the positive values for which parabola and hyperbola intersect ["la jonction de ces deux lieux"]. He devotes the following paragraphs to the construction of such points, if possible.

At this point, and before delving into the above-mentioned construction, take note of the use of a geometric via rather than proceeding to solve the system of equations. From this algebraic point of view, it is immediate to deduce, by replacing $y = \frac{b^2}{x}$ in the equation of the parabola, that the value x must verify the equation of third degree $x^3 = a^2 x - 2ab^2$; with our current mathematical tools (involving differential calculus) it is a simple matter to see that three cases can occur depending on whether b^2 is greater than, or equal to, or less than $a^2 \cdot \sqrt{\frac{1}{27}}$:

(i) if $b^2 < \frac{a^2}{\sqrt{27}}$, there are two positive roots for x;

(ii) if the equality holds we obtain $x = \frac{a}{\sqrt{3}}$, and consequently $y = \frac{b^2}{x} = \frac{a}{3}$, which gives an equilateral triangle;

(iii) if $b^2 > \frac{a^2}{\sqrt{27}}$ there is no solution of the problem.

Obviously, with the lack of differential techniques (notice that the manuscript may be dated 1674), Ozanam is itself restricted to a geo-

---

[60]Latus rectum or the lenght of the chord of the parabola passing for the focus (here, in our current Cartesian coordinates, the point $(0,0)$) and parallel to the directrix (here, the line $y = a$). Notice that its vertex is $(0, \frac{a}{2})$.

[61]"De la jonction de ces deux lieux, on tirera la construction suivante".

metric determination of the sought-after values for the sides and the perpendicular of triangle ABC. Let us explain his construction (see Figure 11):

1. He takes GI as the axis of the parabola HGK.

2. G is the vertex of the parabola,[62] and since the parameter (or latus rectum) is $2a(= AE)$, he marks the line GA having length[63] $\frac{1}{4}(2a) = \frac{a}{2}$, being A the focus of the parabola.

3. From A, he draws the (infinite) line AL perpendicular to the axis GI, and on this segment he chooses the point N holding $AN = b(= AF)$.

4. From the last point N, perpendicular to the line AL is drawn, and on this perpendicular the point O is chosen so that $NO = b(= AF)$.

5. Then, from point A and passing by point O, the hyperbola POQ is traced having the lines AL and AG as its asymptotes.[64]

6. Ozanam realizes that the hyperbola and the parabola intersect at two points, denoted both by C.

7. From each point C, it is considered the perpendicular to the axis AG, so this line determines a segment AD, with D in the axis of the parabola; and immediately it is marked in the axis AG the point B (different from D) such that $DB = DA$.[65]

8. Finally, two (isosceles) triangles ABC are built and each of them has area $b^2(= AF^2)$ and perimeter $2a(= AE)$.

Once the construction is finished, Ozanam continues his development by proving that, indeed, the area and the perimeter of ABC are b^2 and $2a$, respectively. We explain his *demonstration*:

---

[62] In Cartesian coordinates, we see that $G = (0, \frac{a}{2})$.

[63] Thus, in Cartesian coordinates we would have that A is the focus of the parabola, with $A = (0,0)$.

[64] Notice that, then, in Cartesian coordinates, since the point $O(b, b)$ cuts the hyperbola, we have $xy = b^2$.

[65] In Figure 11, DC and $DB = DA$ are denoted by x and y, respectively.

(I) For the area, Ozanam argues that it is equal to the area of the rectangle ADC, and according to the property of the hyperbola, this last area coincides with the area of the rectangle ANO, namely, b^2, which ends the proof for the area.

(II) Concerning the perimeter, he affirms that "par la nature de la parabole" [from the nature of the parabola] it holds[66] $AE \cdot GD = CD^2$. Next, he uses the facts[67] $AG = \frac{1}{4}AE$, $GD = AG - AD = \frac{1}{4}AE - AD$, and $CD^2 = AC^2 - AD^2$ by Pythagorean Theorem. By replacing these quantities in the proportion determined by the nature of the parabola, he obtains

$$\frac{1}{4}AE^2 - AD \cdot AE = AC^2 - AD^2$$

and by antithesis (namely, transposition of terms),

$$\frac{1}{4}AE^2 - AD \cdot AE + AD^2 = AC^2.$$

Since the first part of the equality corresponds to the square of $\frac{1}{2}AE - AD$ (to justify it, Ozanam appeals to the fourth proposition of Book II of Euclid's *Elements*[68]), he deduces that $\frac{1}{2}AE = AC + AD$. Thus,

$$AE = 2AC + 2AD = AC + CB + 2AD = AC + CB + AB,$$

as desired. This ends the proof for the perimeter of the triangle ABC.

---

[66]Ozanam uses the following notation AE, $GD \sim CD_q$, where we can perceive the use of cossist notation, probably a vestige from Viète's legacy. On the other hand, note that, in our current language, equality is nothing more than saying that point (x, y) satisfies the equation of the parabola, since if we take into account that $AE = 2a$, $GD = \frac{a}{2} - y$, and $x = CD$, then we have $2a(\frac{a}{2} - y) = x^2$, or $a^2 - 2ay = x^2$. Nevertheless, in absence of Cartesian notation and algebraic language, this property was well known. The construction and properties of conic curves were discussed by Ozanam in his "Traité des lignes du premier genre expliquées par une methode nouvelle & facile", [Ozanam, 1687(b)].

[67]Recall the definition of a parabola as the locus whose points verify that its distance to the focus (here A, the origin $(0,0)$) is equal to its distance to the directrix (here $y = a$); as the vertex G is $(0, \frac{a}{2})$, we deduce that the distance from G to A is exactly $\frac{a}{2}$, a quarter of the latus rectum.

[68]To wit: If a straight line AB is cut, at random, at point C, the rectangle contained by AB and BC, plus the rectangle contained by BA and AC, is equal to the square on AB.

Obviously, Ozanam has considered that the intersection of the two loci consists of two points. To finish his study, he explains what happens when *the intersection is reduced to a single point* (see Figure 12).

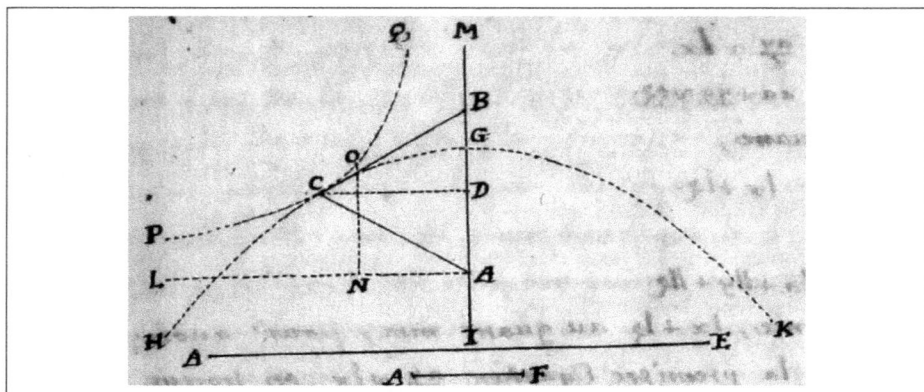

Figure 12: The picture by Ozanam for the case in which the intersection is reduced to a point C, [Ozanam, 1674?, Livre VI, p. 429].

(1) If the intersection between the parabola and the hyperbola is a single point C, Ozanam notes that the two loci will have a common tangent from C.

(2) Then, according to the nature of the parabola,[69] it holds that $GB = GD$.

(3) Now, according to the nature of the hyperbola,[70] he establishes that $AD = DB$.

(4) From the above steps, it is clear that $DB = AD = 2GD = 2GB$ and also $AG = 3BG$.

(5) Taking into account that $AG = \frac{a}{2}$, from the last point he deduces that $BG = \frac{a}{6}$.

---

[69] He employs the property stating that if the tangent of a parabola is traced from a point C and this tangent cuts the parabola's axis MI at a point B, then $GB = GD$, where G is the vertex of the parabola and D is the point from the axis obtained by tracing the perpendicular from C to the axis.

[70] In this case, given a point C on the hyperbola, trace its tangent and assume that TCt is the segment intercepted by the asymptotes AM and AL; then, the segment is divided in two equal parts from the point C. Therefore, by similarity of triangles, it is easily seen that $AD = BD$.

(6) Now, from the property of the parabola stating that $AE \cdot GD = CD^2$, with $AE = 2a$ and $GD = \frac{a}{6}$, he obtains[71] $CD = \sqrt{\frac{a^2}{3}}$; and, on the other hand, $AD = \frac{a}{3}$ since $AD = 2GD$.

(7) Being C a point of the hyperbola, Ozanam observes that it must be $CD \cdot AD = b^2$, or $\frac{a}{\sqrt{3}} \cdot \frac{c}{3} = b^2$, that is, $a^4 = 27b^4$, or equivalently $a = b\sqrt{\sqrt{27}}$.

(8) Then, he proceeds to give the last equality in terms of a continuous proportion: since it is $a, b :: \sqrt{\sqrt{27}}, 1$, he can write $\frac{1}{2}AE, AF :: \sqrt{\sqrt{27}}, 1$. Even, by duplicating the ("En doublant les antecedens") antecedents in the proportion, he finds

$$AE, AF :: \sqrt{\sqrt{432}}, 1$$

(remember that :: is a proportion, that is, $\frac{AE}{AF} = \frac{\sqrt{\sqrt{432}}}{1}$).

He concludes the analysis of this case of unique intersection by remarking that, since the ratio of the given lines AE and AF is equal to this other ratio $\sqrt{\sqrt{432}}, 1$, when the hyperbola touches the parabola, this ratio AE, AF must be less[72] than the other ratio in order to have two isosceles triangles with the same area and perimeter. Finally, Ozanam reminds us that it is possible to solve the problem by reducing it to the intersection of a circle and a parabola.

> This Problem can also be solved very conveniently by the intersection of a circle and a parabola, but the proof will not be so simple, nor the determination so easy: & although these two Loci can be cut into three points, however, there will only be two that will be used, namely the two that will meet on the same side, because the third will give a false root, which cannot be of any use here.[73]

---

[71] As a reminiscence of Descartes' notation, Ozanam writes $\sqrt{\frac{aa}{3}}$, that is, he expresses the square by aa.

[72] Notice that Ozanam asserts that the ratio must be less than the other ratio ("elle doit être moindre"), when the ratio must be greater than the other ratio, maybe it was a typo.

[73] "Ce Probleme se peut aussy resoudre tres commodément par l'intersection d'un

At this point, the reader would have to be aware of his book [Ozanam, 1687(a)] in order to be able to carry out this reduction, but nothing is said about it; Ozanam only mentions the fact that, in this new geometric setting, it is possible to find three points, but one of them will be false and will not serve for the resolution of the problem.

Finally, a little remark. It is interesting to note that this problem also appeared in the aforementioned *Traité de la construction des equations* [Ozanam, 1687(a), pp. 20-26]. In this case, however, no mention is given to the possibility of reducing the problem to the intersection of a suitable circle and parabola.

4   Conclusions

We can certainly conclude that Ozanam was an outstanding mathematician, judging by the large number of his works and the numerous reprints and reissues of his books, even after his death. The breadth of mathematical topics he dealt with also demonstrates his mastery of the mathematics of his time, which he used to teach and disseminate. Ozanam had a solid mathematical background based on the Jesuit tradition, probably rooted in his relationship with mathematicians such as Deschales or De Billy.

Ozanam was in contact with important mathematicians of his time and received well-deserved praise from his contemporaries, such as Leibniz, De Billy and Fontenelle, as we have reinforced with different quotations. In particular, it is worth mentioning the words of Leibniz praising Ozanam's treatise on Algebra and his analytical method that allowed him to solve problems "which could not be solved by the Diophantine or any algebraic method known hitherto".

It also seems evident that his mastery of an advanced algebraic language, jointly with the use of his analytical method, allowed him to develop general solutions as well as a rigorous treatment of the questions he addressed. Moreover, as we have shown in the study of Ques-

---

cercle & d'une parabole, mais la demonstration ne sera pas si simple, ny la determination si aisée: & bien que ces deux Lieux se puissent couper en trois poins, il n'y en n'aura pourtant que deux qui serviront, savoir les deux qui se rencontreront d'un même côté, parce que le troisieme donnera une racine fausse, qui ne peut icy être d'aucun usage", [Ozanam, 1674?, Livre VI, p. 429].

tion XVIII of Book VI and its ten added problems of his *Diophantus*, he not only disseminated mathematics with an advanced symbolic language, but also added new problems, improved the solution of others, and even posed and investigated new questions. Ozanam followed Bachet in writing his *Diophantus* in terms of the structure and distribution of the results of the treatise. However, we observe clear differences in the procedure of solving the problems by each author. While Bachet used a rhetorical language with a marked geometrical point of view based on Euclid's *Elements*, Ozanam used an analytical method together with an advanced symbolic language that allowed him to obtain general solutions, including their algebraic characterization. Thus, his expertise in algebraic methods have also been revealed in the analysis of his solutions, not only by the way he calculated them but also by the way he studied such solutions with possible alternatives or generalizations.

His mathematical skill and profound insight into mathematics are evident in his geometrical ability to handle proportions, such as the bisector theorem, and in his knowledge of the properties of conics, as we have seen in the case of parabolas and hyperbolas, but all sifted through an algebraic language and an analytical method, clearly influenced by Descartes' procedures, that abandon Euclid's classical treatment of geometry. In the questions that we have analyzed in this work, we observe how, when considering a geometric problem, he translates it into algebraic terms, assigning unknowns and parameters to the different parts of the figure, from which he obtains general solutions. In some cases, the algebraic solution of the problem involves the relationship between equations and loci. Then, once the problem is solved, it returns to the geometric field to build the solutions.

Descartes' influence can also be seen in the use of a similar notation as well as the identification of equations with curves or loci, associating canonical equations for conic curves. Even, we have found vestiges from Viète, such as homogeneous law and the notation (CD_q) for the square of a segment. Nevertheless, Ozanam's procedures are original, as Leibniz himself stated when referring to his algebra, which he found "much better [...] than most of the ones we have seen for some time, which only copy Descartes and his commentators". The process of algebrization thus finds in Ozanam's work a significant step in its consolidation during the late 17th and early 18th centuries. In fact, the use of his symbolic language makes the reading of his work easily understandable for a current reader.

We are in the presence of a giant of mathematics because of his depth, his mastery of concepts and ideas, his skill in the use of algebraic language within his analytical procedures, the relationship between arithmetic, algebraic, geometric or applied problems and, finally, because of his extraordinary work, which covers all fields of mathematics. Perhaps these adjectives would have been more pronounced if his *Diophantus*, which remained in manuscript form without being printed, had been available to the general public. In this paper, we conjecture a reason for its non-publication, namely the unforeseen death of Jean-Baptiste Colbert, a prominent member of the court of Louis XIV, who, in Ozanam's own words, wished to print Ozanam's *Diophantus*.

Ozanam's work was widely disseminated and, therefore widely used and accepted as a reference in the study and learning of mathematics in the 17th and 18th centuries.

Acknowledgement

The authors wish to express their sincere gratitude to Davide Crippa, whose comments, suggestions and indications have greatly improved this work.

References

[Bachet, 1621] C.G. Bachet. *Diophanti Alexandrini Arithmeticorum Libri sex, et De Numeris Multangulis Liber unus.* Sumptibus Hieronymi Drovart, Paris, 1621.

[Bos, 2001] H. J. M. Bos. *Redefining Geometrical Exactness, Sources and Studies in the History of Mathematics and Physical Sciences.* Springer-Verlag, New York, 2001.

[Cajori, 1993] F. Cajori. *A History of mathematical notations* (two volumes). Dover publications INC., Mineola, NY, 1993.

[Càndito, 2016] C. Càndito. Jacques Ozanam (1640-1718). In M. Cigola, editor, *Distinguished Figures in Descriptive Geometry and Its Applications for Mechanism Science*, pages 223-248. Springer International Publishing Switzerland, 2016.

[Cassinet, 1986] J. Cassinet. Le Traité des simples, des doubles et des triples égalités, de Jacques Ozanam dans le manuscrit autographe perdu puis retrouvé de la Bibliothèque du Chancelier d'Aguesseau. *Cahiers du séminaire de Toulouse*, 8: 69-96. 1986.

[Clément, 1846] P. Clément, *Histoire de la vie et de l'administration de Colbert*. Guillaumin Libraire, Paris, 1846.

[Crippa, 2014] D. Crippa. *Impossibility results: From Geometry to Analysis*. Université Paris Diderot: Phd Dissertation. 2014.
Available: https://hal.science/tel-01098493

[Crippa, 2019] D. Crippa. *The Impossibility of Squaring the Circle in the 17th Century. A Debate Among Gregory, Huygens and Leibniz*, Birkhäuser, Cham, 2019.

[Descartes, 1637] R. Descartes. *Discours de la méthode. Pour bien conduire sa raison, et chercher la vérité dans les sciences*. Ed. Ian Maire, Leyde, 1637.

[Fontenelle, 1719] B. Fontenelle. Eloge de M. Ozanam. *Histoire de la Académie Royale des Sciences, Année MDCCXVII*, pages 86-92. Imprimerie Royale, Paris, 1719.

[Foucher, 1854] A. Foucher de Careil (ed.). *Lettres et opuscules inédits de Leibniz. Précédés d'une introduction*. Librairie Philosophique de Ladrange, Paris, 1854.

[Hall, 1973] A.R. Hall; M.B. Hall (eds. and transl.). *The Correspondence of Henry Oldenburg*. Vol. IX, pp. 488-498. The University of Wisconsin Press, Madison, 1973.

[Hall, 1977] A.R. Hall; M.B. Hall (eds. and transl.). *The Correspondence of Henry Oldenburg*. Vol. XI, pp. 98-108. Mansell, London, 1977.

[Gómez et al., 2021] F. Gómez-García, P.J. Herrero-Piñeyro, A. Linero Bas, M.R. Massa-Esteve, A. Mellado Romero. The six books of Diophantus' Arithmetic increased and reduced to specious: the lost manuscript of Jacques Ozanam (1640–1718). *Arch. Hist. Exact Sci.* 75: 557–611, 2021.

[Heath, 1910] T.L. Heath. *Diophantus of Alexandria. A Study in the History of Greek Algebra*, sec. Ed. Cambridge University Press, Cambridge, 1910.

[Journal, 1692] Extrait d'une lettre de M. de Leibniz à Mr. Foucher Chanoine de Dijon, sur quelques axiomes de Philosophie. *Le Journal des Savants*. Chez Jean Cusson pp. 247-248. Paris, 1692.

[Journal, 1703] [...] Quelques autres Remarques sur le même Traité d'Algebre à l'occasion de celle de M. Leibnitz. *Le Journal des Savants*. Chez Jean Cusson pp. 364-368. Paris, 1703.

[Leibniz, 1703] G.W. Leibniz, Remarque de M. Leibniz, sur un endroit des nouveaux Elements d'Algebre de M. Ozanam. *Le Journal des Savants.* Chez Jean Cusson pp. 362-364. Paris, 1703.

[Leibniz, 1993] G.W. Leibniz. *De quadratura arithmetica circuli ellipseos et hyperbolae cujus corollarium est trigonometria sine tabulis* (edited by E. Knobloch), Vandenhoeck und Ruprecht, Göttingen, 1993.

[Mahoney, 1980] M.S. Mahoney. The beginnings of algebraic thought in the seventeenth century. In S. Gaukroger (ed.), *Descartes' Philosophy, Mathematics and Physics*, pages 141–156. Harvester Press, Brighton, 1980.

[Mancosu, 1996] P. Mancosu. *Philosophy of Mathematics and Mathematical Practice in the Seventeenth Century.* Oxford University Press, Oxford, 1996.

[Massa-Esteve, 2001] M.R. Massa Esteve. Las relaciones entre el álgebra y la geometría en el Siglo XVII. *Llull*, 24: 705–725, 2001.

[Massa-Esteve, 2012] M.R. Massa Esteve. The role of symbolic language in the transformation of mathematics. *Philosophica*, 87: 153–193, 2012.

[Mellado Romero, 2022] A. Mellado Romero. *La influencia del Cursus Mathematicus de Hérigone en la algebrización de la matemática.* PhD Dissertation, University of Murcia, Spain, 2022.
Available: http://hdl.handle.net/10201/123566

[Meskens, 2010] A. Meskens. *Travelling Mathematics. The Fate of Diophantos' Arithmetic.* Birhhäuser, Basel, 2010.

[Montucla, 1758] J.E. Montucla. *Histoire des matématiques.* Chez CH. Ant. Jombert, Paris, 1758.

[Niceron, 1739] R.P. Niceron. *Mémoires pour servir à l'histoire des hommes illustres dans la république des lettres XL.* Chez Briasson, Paris, 1739.

[Ozanam, 1673] J. Ozanam. *Traité de gnomonique, ou de la construction des cadrans sur toute sorte de plans.* Claude Cramoisy, Paris, 1673.

[Ozanam, 1674?] J. Ozanam. *Les six livres de l'arithmetique de Diophante d'Alexandrie augmentez & reduits a la specieuse.* Kept by the *Accademia delle Scienze di Torino* (with the shelf mark MSR. 0144-0145). Unpublished manuscript (approx. 1674).

[Ozanam, 1684] J. Ozanam. *La Géométrie pratique, contenant la trigonométrie théorique et pratique, la longimétrie, la planimétrie et la stéréotomie.* L'auteur et Etienne Michallet, Paris, 1684.

[Ozanam, 1687(a)] J. Ozanam. *Traité de la construction des equations pour la solution des problemes indeterminez.* Chez Estienne Michallet, Paris, 1687.

[Ozanam, 1687(b)] J. Ozanam. *Traité des lignes du premier genre expliquées par une methode nouvelle & facile.* Chez Estienne Michallet, Paris, 1687.

[Ozanam, 1688] J. Ozanam. *L'Usage du Compas de Proportion expliqué et dé-*

montré d'une manière courte et facile et augmenté d'un Traité de la division des Champs. Estienne Michallet, Paris, 1688.

[Ozanam, 1688(b)] J. Ozanam. *Usage de l'instrument universel. Pour resoudre promtement & tres-exactement tous les Problemes de la Geometrie Pratique sans aucun calcul.* Estienne Michallet, Paris, 1688.

[Ozanam, 1693] J. Ozanam. *Cours de mathématique, qui comprend toutes les parties de cette science les plus utiles et les plus nécessaires à un homme de guerre, et a tous ceux qui se veulent perfectionner dans les mathématiques,* 5 volumes. Jean Jombert, Paris, 1693.

[Ozanam, 1694] J. Ozanam. *Récréations mathématiques et physiques, qui contiennent plusieurs problèmes d'arithmétique, de géométrie, de musique, d'optique, de gnomonique, de cosmographie, de mécanique, de pyrotechnique, et de physique. Avec un traité des horloges élémentaires,* 2 vol. Ed. Jean Jombert, Paris, 1694.

[Ozanam, 1699] J. Ozanam. *Traité de l'arpentage et du toisé, ou Méthode facile pour arpenter ou mesurer toutes sortes de supeficies, et pour toiser exactement la maçonnerie, les vidanges des terres, et tous les autres corps, dont on peut avoir besoin dans la pratique; avec le toise du bois de charpente selon la Coutume de Paris, et un traité de la séparation des terres.* Jean Jombert, Paris, 1699.

[Ozanam, 1702] J. Ozanam. *Nouveaux elements d'Algebre, ou Principes Generaux, Pour resoudre toutes sortes de Problémes de Mathematique,* 2 vol. Chez George Gallet, Amsterdam, 1702.

[Ozanam, 1712] J. Ozanam. *Cursus Mathematicus: Or, A Compleat Course Of The Mathematicks. In Five Volumes,* 5 vol. Ed. John Nicholson, London, 1712.

[Ozanam-Montucla 1758] J. Ozanam. *Récréations mathématiques et physiques,* ed. J. E. Montucla (2 vol.). Ed. Cl. Ant. Jombert, Paris, 1758.

[Ozanam, 1851] J. Ozanam. *Recreations in science and natural philosophy,* Rev. Edward Riddle. William Tegg and Co., London, 1851.

[Pelay, 2011] N. Pelay. Jeu et apprentissages mathématiques: élaboration du concept de contrat didactique et ludique en contexte d'animation scientifique. Anexe E1. Université Claude Bernard - Lyon I, 2011. Available: https://tel.archives-ouvertes.fr/tel-00665076

[Salomon-Bayet, 1978] C. Salomon-Bayet. Les académies scientifiques: Leibniz et l'Académie Royale des Sciences (1672–1676). In: *Leibniz in Paris (1672-1676: Symposion de la G. W. Leibniz-Gesellschaft (Hannover) et du Centre National de la Recherche Scientifique (Paris) à Chantilly (France) de 14 au 18 november 1976.* Steiner, Wiesbaden, 1978.

[Schaaf, 1990] W.L. Schaaf. Biography. In *Dictionary of Scientific Biography*. Charles Scribner's Sons. New York, 1970-1990.

[Serfati, 2010] M. Serfati. Symbolic revolution, scientific revolution: mathematical and philosophical aspects. In Heeffer, A. & Van Dyck, M. (eds.). *Philosophical Aspects of Symbolical Reasoning in Early Modern Mathematics*, Vol. 26. College Publications, London, 2010.

[Sergescu, 1936] P. Sergescu. Les Mathématiques dans le Journal des Savants: Première période 1666-1701. *Osiris*, 1: 568-583,1936.

[Stroup, 1987] A. Stroup. Royal Funding of the Parisian Academie Royale des Sciences during the 1690s. *Transactions of the American Philosophical Society*, 77: 1-167, 1987.

[Stroup, 1996] A. Stroup. Christian Huygens et l'Académie Royale des Sciences, *Vie Sci*, 13: 333-341, 1996.

[Viète, 1591] F. Viète. *In Artem Analyticen Isagoge: seorsim excussa ab opere restitutae mathematicae analyseos seu algebra nova*. Ed. Jacques Mettayer, Toronis, 1591.

[Wallis, 1699] J. Wallis. *Operum Mathematicorum*. Vol. 3, 618-619 (1699). E Theatro Sheldoniano, Oxford.

The impact of analysis on the understanding of irrational quantities. Legendre's contribution in context

Eduardo Dorrego López

I.E.S. Afonso X O Sabio, A Barcala, A Coruña, Spain

edorregolopez@gmail.com

Abstract

This chapter will examine how the change of paradigm occurred with the introduction of analysis changed the prevalent understanding of irrational quantities. Due to their being particularly important, we present a brief sketch of Euler's and Lambert's contribution to the comprehension of irrational quantities: Euler proved for the first time the irrationality of e and Lambert the irrationality of π. As a case study, we focus on Legendre's proof of the irrationality of π and π^2 —the former being conspicuously shorter than Lambert's—, presenting to the reader an easy-to-follow analysis of his proof, and a contextualized study of these remarkable results included in his *Élements de Géometrie* (1794).

1 Introduction

As far as mathematics is concerned, the development of analysis may be said to have been the main line of research in the 18th century. However, this was not the only mathematician's preoccupation. There were secondary subject matters investigated at that time, such as the attempts to elucidate the status of negative and infinitesimal quantities, the challenges derived from the apparent appearance of new geometries, or number-theory-related issues, just to mention a few.

The main goal of this chapter is to focus on one of these 18th-century secondary lines of research, namely, proofs of irrationality. The first section is intended to explain how the change towards analytic methods helped to put the focus on some "special irrationals", particularly on π, giving rise to a new, although not yet well-understood, kind of irrationals: the so-called transcendental quantities.

In the second section, we present what was to become the key analytical tool used in order to deal with special irrationals: continued fractions. In doing so, some mathematicians who were responsible for having developed this tool in its first stage are briefly analysed along with the role they played, such as Euclid, Cataldi, Bombelli, Brouncker or Wallis. The main part of this section, however, is devoted to Euler and Lambert, who proved for the first time the irrationality of e and π, respectively.

The main contribution of this chapter is included in the third and last section, which is devoted to Legendre and his proof of the irrationality of π and π^2 and in which the reader will find a detailed explanation of the steps taken by Legendre. After the presentation of Legendre's proof, we include in the concluding remarks some reflections about the possible influence of Euler and Lambert, as well as a comparison between Lambert's and Legendre's proofs. To some extent, Legendre's proof can be considered as the follow-up to what Lambert have done, but, as it will be argued, there are also deep methodological and technical differences.

2 The role played by analytic methods in the awareness of special irrationals

The algebraization of mathematics, the process which we investigate in the present book, had a significant impact on how the various branches of mathematics were perceived in the 18th century. This impact is emphasized by Bos in connection with geometry, bringing into focus the transformations undergone by this classic discipline after the introduction of analytic methods:

> It was from this field [the investigation of curves by means of finite and infinitesimal analysis] that, in the period 1650–

1750, infinitesimal analysis gradually emancipated itself as a separate mathematical discipline, independent of the geometrical imagery of coordinates, curves, quadratures, and tangents, and with its own subject matter, namely, analytical expressions and, later, functions. This process of emancipation, which might be called the degeometrization of analysis, constituted the principal dynamics within the area of mathematical activities around the investigation of curves by means of finite and infinitesimal analysis. It was strongly interrelated with the changing ideas on the interpretation of exactness with respect to construction and representation.[1]

But the consequences that this process had for mathematics extended far beyond geometry, opening new lines of research within well-established areas of mathematics, or creating new frameworks favourable to the flourishing of new concepts. As an example of the first case, we might cite the attempts to use algebra to turn classical theory of numbers into a general discipline, in such a way that it would be possible not only to extract properties in specific cases, but also to formulate general theorems.[2] And as an example of the second case, we might cite the application of analytic methods to the elucidation of special irrationals, which is what this chapter will be focusing on.[3]

Although it is difficult to advance this claim with absolute certainty, the shift to analytical mathematics was likely responsible for the emergence of a clear awareness of the existence of another kind of irrational quantity besides common ones. Over the centuries, irrationals were used predominantly in connection with the resolution of algebraic equations. By means of concrete methods, mathematicians were able to

---

[1][Bos, 2001, p. 10].

[2]See [Fraser, 1990, p. 246], for further details. Fraser cites the name of Malebranche as one of the prominent authors who followed this current in the first stages of this process of algebraization, as well as Newton, who used the term "universal arithmetic" in connection with this idea (it is not difficult to find textbooks throughout this century, and beyond, bearing in their titles a reference to this programme, for example Cristian Kramp's *Arithmétique Universelle*, 1801).

[3]What is meant by "special irrational' will become clear as the reader proceeds through the present chapter, although we can anticipate that it refers to quantities like π or e (transcendental quantities).

arrive at some of their solutions which, after the introduction of decimals, were characterized by having an infinite amount of non-periodic decimal places (in modern terms),[4] which was actually what motivated the term *surds*.[5]

However, the surge in the development of algebra turned this apparently intangible property into a more accessible way to manage these solutions. The algebraic language allowed these unnamed quantities to be described using a finite combination of basic algebraic operations. In this way, irrational quantities re-emerge under a new form: $\sqrt{2}$, $\sqrt[3]{3 + \sqrt{2}}$, etc. They now also bore the appropriate name of *radicals* at large.

On the other hand, the application of these new techniques to old geometric problems brought to light a new, hidden aspect of well-known quantities, the most notorious among these problems being the quadrature of the circle. During its long existence, there have been three strategies to tackle this, depending on whether geometry, arithmetic or analysis is used. For a long time the standard approach was to address this problem using classical geometry alone, so it was not easy to let both arithmetic and algebra begin to play the role of supporting strategies.[6] Once analytic methods began to be widely applied, the focus within the circle-squaring problem slowly shifted from a geometric to an analytic aspect.

Very much in keeping with this new analytic line of research was that of James Gregory (1638–1675), who published a treatise in 1667, *Vera circuli et hyperbolae quadratura*, that kept leading mathematicians occupied in a controversy about the attempts of squaring the circle. Gregory was able to reduce the quadrature problem (that of squaring geometrically the sector of a conic) to expressing a certain magnitude,

---

[4] As is well known, this feature of irrational quantities is what led Stifel to describe them in the following terms in his *Arithmetica integra*, 1544: "just as an infinite number is no number, so an irrational number in not a true number, because it is so to speak concealed under a fog of infinity." (Stifel, quoted in [Ebbinghaus et al., 1988, p. 33]).

[5] According to [Smith, 1958, p. 252], it was the distinction made by Al-Khowârizmî between "audible" and "inaudible" numbers that gave rise to the term "surd" (deaf, mute). As far as we know, this term was used for the first time in Europe by Gherardo of Cremona around 1150.

[6] As for the tensions among arithmetic, algebra and geometry in the 16th and 17th centuries see [Bos, 2001, chapters 6, 7, and 8].

the area of this sector, by means of addition, subtraction, multiplication, division and extraction of roots of kth order, and he even came to believe, at first, that the problem might actually be capable of being solved in this way. As in many other cases, though, the obstacles finally convinced him of quite the opposite, making his *Vera circuli* a monument dedicated to the impossibility results. With regard to the circle-squaring problem, he concluded its impossibility in a corollary, so clear enough to Gregory was not only the non-rationality of π, but also that this quantity could not be expressed by means of a finite amount of algebraic operations.[7]

Gregory's treatise was not spared criticism by other mathematicians, but it helped to put the circle-squaring problem into a new (analytical) light and consequently to put its arithmetical twin π, into focus. In his *Arithmetica infinitorum*, published in 1656, John Wallis (1616–1703) placed the focus on seeking the quadrature of certain curves, with special attention paid to the circumference. Although the path followed by Wallis is based on an arithmetic approach instead of an algebraic one as in other cases —using divergent series and numerical interpolations in order to get the inverse of the ratio between the circle and its circumscribed square—,[8] his method let him to obtain this ratio through an analytic formula, such as the one he writes in Proposition 191:

$$\frac{3 \times 3 \times 5 \times 5 \times 7 \times 7 \times \cdots}{2 \times 4 \times 4 \times 6 \times 6 \times 8 \times \cdots}.$$

He did not represent this expression in this way, but the product is nothing but $\frac{4}{\pi}$, and, as he proved, it is possible to get approximations as accurate as one wants, which, along with the applicability of the usual arithmetical rules, led him to consider it "as valid as any other commonly accepted number".[9] However, as far as the subject of this chapter is concerned, the most interesting thing is the conclusion that Wallis arrived at concerning this number: drawing on the aforementioned infinite product, all seem to indicate that this number cannot be considered as a common irrational quantity:

---

[7]An exhaustive study of James Gregory along with quadrature problems in the 17th century is to be found in [Crippa, 2014] and [Crippa, 2019].

[8]See [Wallis, 1656/2004, pp. xiv, xviii].

[9][Wallis, 1656/2004, p. xxv].

And indeed I am inclined to believe (what from the beginning I suspected) that this ratio we seek is such that it cannot be forced out in numbers according to any method of notation so far accepted, not even by surds (of the kind implied by Van Schooten in connection with the roots of certain cubic equations, in his *Appendix* to the treatise *On a complete description of conic sections*, or in the thinking of Viète, Descartes and others) so that it seems necessary to introduce another method of explaining a ratio of this kind, than by true numbers or even by the accepted means of surds.[10]

In fact, this "another method" is precisely the product obtained in Proposition 191, a product that, according to his analysis, cannot be expressed algebraically (that is, using a finite combination of algebraic operations), which means that it is impossible to frame it within the field of known numbers, be these rationals ("true numbers") or not ("by the accepted means of surds").[11]

It seems that the discussions on the nature of π were greatly reinvigorated by the results obtained upon applying these mathematical tools to its twin geometric problem. The apparent impossibility of being reduced to a finite combination of usual algebraic operations nourished the idea of the existence of another group of quantities essentially different from the common irrationals.

As a matter of fact, it was also in this epoch, in 1673, and from the hand of another distinguished mathematician, Gottfried Wilhelm Leibniz (1646–1716), that the concept of transcendence appears in the field of mathematics precisely in order to refer to this kind of infinite

---

[10][Wallis, 1656/2004, p. 161]. In the original: "et quidem proclivis sum ut credam (quod et ab initio suspicatus sum,) rationem illam quam quaerimus talem esse quae non poterit numeris exprimi iuxta ullum adhuc receptum notationis modum, ne quidem per latera surda; quale quid invenit Schotenius, de radicibus aequsationum quaerendam cubicarum, in ipsius Appendice ad tractatum de Organica Conicarum Sectionum descriptione, idq. ad mentem Vietaei, Cartesii, & aliorum :) ut necesse videatur alium ejusmodi rationem explicandi modum introducere, quam vel per numeros veros, vel etiam per latera surda" (at p. 174).

[11]In fact, this led Wallis to introduce a new notation for his infinite product (see [Wallis, 1656/2004, p. 163]). Stedall comments that Wallis might have taken this notation from Pierre Hérigone (1580–1643), on account of Wallis referring to him in his "*A Treatise of Algebra*, p. 128 (London, 1685) as a source of new notation" ([Stedall, 2000, p. 314, note 19]).

expressions.[12] It must be said, however, that the range of applicability of this concept was much wider than the current one: curves, figures, problems, equations, quantities, or numbers were susceptible to carrying this qualification, making it difficult to obtain a concise understanding of its meaning. As for π, Leibniz, who obtained another analytic expression after using the analytic machinery on the circle-squaring problem (although in his case it was an infinite series):[13]

$$\frac{\pi}{4} = 1 - \frac{1}{3} + \frac{1}{5} - \frac{1}{7} + \frac{1}{9} - \cdots,$$

seems to have been quite clear that its nature was also more restrictive than rational numbers' or surds'.[14]

What this analytic approach provided for irrational quantities in these first steps towards an algebraization of mathematics was, therefore, a new framework that made it possible to recognize clear differences among them: expressible irrational quantities (usually called *radicals* or *surds*) vs. non-expressible irrational quantities (usually called *transcendental*). As the reader will have noted, this 17th-century terminology is at odds with the one currently used, at least when we refer to irrational numbers. In any case, it does not fall within the scope of this chapter to delve into how this classification was to turn into the modern one —roots of equations (*algebraic* numbers) vs non-roots of equations (*transcendental* numbers)—, but just to bring out how this analytic movement established the bases for a manageable and adequate way of distinguishing between quantities (in this first stage by using the analytic expressibility as a yardstick).[15]

---

[12]See [Crippa, 2014, p. 418, note 134]; or [Knobloch, 2006, p. 122]. For an exhaustive study of the concept of transcendence in Leibniz see [Serfati, 2018].

[13]Leibniz found this series in 1673 based on the arctang series. J. Gregory and Nilakantha found independently this very expression [Berggren et al., 1997, p. 92]).

[14]As for the concept of number in Leibniz, see [Sereda, 2017].

[15]Although much work remains to be done, my dissertation represents a first step toward such an in-depth investigation ([Dorrego López, 2021]).

3 Continued fractions as the engine of irrationality

Among the analytic tools in the process of development, so-called continued fractions would prove to be enormously useful in mathematics, not only, but particularly, for irrational-related issues.

The underlying idea can be traced back to Euclid and his algorithm for computing the greatest common divisor between two segments (Prop. 2, Book VII). Brezinski, in his standard book on the history of continued fractions, summarizes this algorithm as follows:

> Euclid measured the greatest segment with the smallest one, then the smallest with the remainder, then the first remainder with the new one, and so on. The last non-zero remainder is the g. c. d.[16]

Putting it into a modern language, in order to find the common measure between two different numbers A and B $(A < B)$, one proceeds all the way down by continually dividing the greatest number by the smallest, obtaining the corresponding quotients $(Q, Q', Q'', \cdots)$ and remainders $(R, R', R'', \cdots)$. Piecing all the resultant data together and using the fractional notation, this method yields a continually-fractional result:[17]

$$\frac{A}{B} = \frac{1}{\dfrac{B}{A}} = \cfrac{1}{Q' + \dfrac{R'}{A}} = \cfrac{1}{Q' + \cfrac{1}{\dfrac{A}{R'}}} = \cfrac{1}{Q' + \cfrac{1}{Q'' + \dfrac{R''}{R'}}} = \cfrac{1}{Q' + \cfrac{1}{Q'' + \cfrac{1}{\dfrac{R'}{R''}}}} = \cdots$$

In any case, as is appropriately pointed out by Brezinski on the same page: "Of course, Euclid did not present his algorithm in that way, nor did he use continued fractions".

---

[16]This method was called by Greeks "anthypharesis" (for a more detailed explanation of Euclid's algorithm, I forward the interested reader directly to [Heath, 1908, p. 298]). As for the history of continued fractions, I rely heavily on [Brezinski, 1991](quote at page 4).

[17]Let us note that this procedure leads to a specific kind of continued fraction, that with all the numerators equal to 1. These continued fractions are called "regular", and will play a key role as far as irrational quantities are concerned.

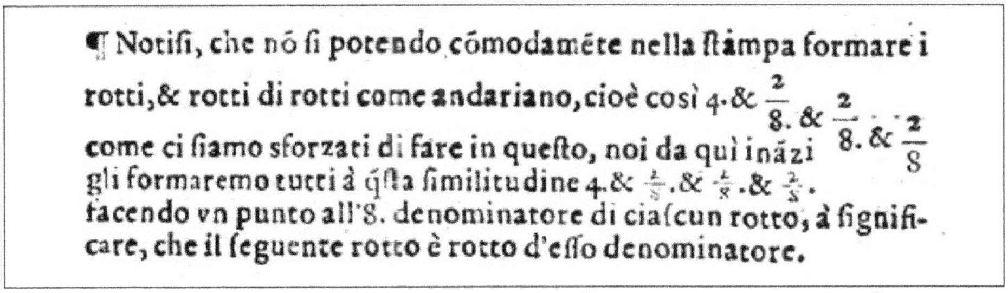

¶ Notiſi, che nó ſi potendo cómodaméte nella ſtámpa formare i rotti, & rotti di rotti come andariano, cioè coſì 4. & $\frac{2}{8}$. & $\frac{2}{8.\&\frac{2}{8}}$ come ci ſiamo sforzati di fare in queſto, noi da qui inázi gli formaremo tutti à q̃ſta ſimilitudine 4. & $\frac{2}{8}$. & $\frac{2}{8}$. & $\frac{2}{8}$. facendo vn punto all'8. denominatore di ciaſcun rotto, à ſignificare, che il ſeguente rotto è rotto d'eſſo denominatore.

Figure 1: Cataldi's notation included in [Cataldi, 1613, p. 70] (image: CC BY-SA. https://www.europeana.eu/item/9200369/webclient_DeliveryManager_pid_88126_custom_att_2_simple_viewer). He also provided, for the sake of convenience in printing, an alternative, more compressed way of writing this expression. The reader can see an English translation of this paragraph in [Brezinski, 1991, p. 66].

The first steps towards the establishment of a theory of continued fractions in the proper sense, were mainly taken by Pietro Cataldi (1548-1626) and John Wallis (1616-1703). Cataldi in his attempts to approximate $\sqrt{18}$, drew on an old recurrent algorithm —used early on in Italy by Rafael Bombelli (1526–1572)—, and used a modern-like notation to express what we would now represent by:[18]

$$4 + \cfrac{2}{8 + \cfrac{2}{8 + \cfrac{2}{8}}}.$$

Besides this, he showed that the fractions obtained by truncating the above-written formula are alternately greater than and smaller than $\sqrt{18}$, and that they approach this value ever more accurately.[19]

---

[18]See Figure 1. Bombelli did not use any modern notation, Pietro Cataldi being the first to develop such a symbolism ([Brezinski, 1991, p. 65]).

[19]"Et così potremo seguire a trovare la ottaua, la nona, la decima, & altre di mano in mano, che più se accostaranno al 18. proposto, l'una in ícarsezza, l'altra in eccesso" (And so we can go on to find the octave, the ninth, the tenth, & others one after the other, which approach more accurately the 18. proposed, one in scarcity, the other in excess) [Cataldi, 1613, p. 71]. For a context-based study of Cataldi's role in the history of continued fractions, the reader may find useful [Favaro, 1875], particularly pp. 55–59. These fractions are what we now appropriately call "convergents".

Wallis was also to take a step forward in this direction in his *Arithmetica Infinitorum*. After obtaining his infinite product for $\frac{4}{\pi}$, he asked William Brouncker (1620–1684), president of the Royal Society, for a more convenient way to express this ratio. His unexpected answer,[20] took the form of a continued fraction:

$$1 + \cfrac{1}{2 + \cfrac{9}{2 + \cfrac{25}{2 + \cfrac{49}{2 + \cdots}}}}.$$

It was actually at this point that Wallis laid the foundations for introducing a new denomination into mathematics in order to represent this kind of analytic expressions, namely "continued fractions", when he referred to them as fractions whose denominators are continually broken.[21] But more important is the fact that Wallis, in his attempts to derive this continued fraction (Brouncker did not explain how he had come up with it), also put the focus on general properties, as for example the alternance in the approximation of the convergents and their increasingly accuracy in the approximation. Furthermore, he also provides a general recurrent formula for these convergents.[22]

In any case, let us point out that these properties brought to light the capacity of continued fractions to approximate irrational quantities, but they did not say anything about the quantities themselves. The seminal work would come almost one century after Wallis' treatise by the hand of Leonhard Euler (1707–1783), and would be successfully applied to the understanding of some special irrationals by this writer himself and, most remarkably, by Johann Heinrich Lambert (1728–1777).

---

[20]"There was nothing in the Arithmetica Infinitorum, or in any known English mathematics up to this point, to prepare the reader for it" ([Stedall, 2000, p. 300]).

[21]"Fractio, quae denominatorem habeat continue fractum" ([Stedall, 2000, p. 300]).

[22]This is included in [Wallis, 1656/2004, pp. 167–178].

3.1 Euler's and Lambert's contributions. A brief sketch

In his *De Fractionibus Continuis Dissertatio*,[23] Euler paved the way for
the use of continued fractions in a more theoretical way. As he noted
at the very beginning, this analytic tool had been hardly used until
that moment, although it showed a noteworthy power to approximate
irrational quantities and, moreover, to express them. This change of
perspective, namely, to put the focus on the more theoretical part of
these tools by studying their shape, led him to formulate a criterion of
irrationality for quantities drawing on regular continued fractions:

> Moreover, every finite fraction whose numerators and de-
> nominators are finite whole numbers may be transformed
> into a continued fraction of this kind which is truncated at
> a finite level. On the other hand, a fraction whose numer-
> ator and denominator are infinitely large numbers (which
> are given for irrational and transcendental quantities) will
> go across to a continued fraction running to Infinity.[24]

Let us clarify that the expression "a fraction whose numerator and denominator are infinitely large numbers", refers to infinite decimal expansions like $\sqrt{2} = 1,41421356\cdots$, which, therefore, would give rise to
fractions with infinitely large numerators and denominators ($\frac{141421356\cdots}{100000000\cdots}$
in this particular case). Besides, as we have mentioned before, we shall
not be delving into the problematic around the terminology, but just
bear in mind that Euler, like Leibniz, Wallis, and the greater part of
his contemporaries (and also later mathematicians actually), was using
"expressibility" as a yardstick to differentiate between inconmensurable
quantities: irrational, radical or surd (finitely expressible, like $\sqrt{2}$) vs
transcendental (non-finitely expressible, like e or π).[25].

---

[23] An essay on continued fractions ([Euler, 1744/1985]). To get a good comprehension of Euler's paper it is advisable to go to [Cretney, 2014].

[24] [Euler, 1744/1985, p. 302]. In the original: "Omnis autem fractio finita, cuius numerator et denominator sunt numeri integri finiti in huiusmodi fractionem continuam
transformatur, quae alicubi abrumpitur; fractio autem cuius numerator et denominator sunt numeri infinite magni, cuiusmodi dantur pro quantitatibus irrationalibus et
transcendentibus, in fractionem vere continuam et in infinitum excurrentem transi-
bit." (at p. 108).

[25] As for the use of the concept of transcendence in Euler see [Petrie, 2012]

Euler did not stop here, but went farther in his analysis, studying "periodic" continued fractions and disclosing their nature as roots of equations. Furthermore, he made a statement that, by using the above-quoted criterion, was to become the first proof of the irrationality of a special irrational:

> Before we proceed to the computation of continued fractions in general whose denominators form an arithmetic progression let us analyze certain transcendental quantities which, when converted into continued fractions, give denominators proceeding in an arithmetic progression. From these examples a straightforward method of computing continued fractions of this sort will arise. Therefore, testing this method by means of logarithms and other transcendental quantities, I have found that the number whose natural logarithm is 1, and its powers, lead to continued fractions of this kind.[26]

Euler provided this continued fraction:[27]

$$e = 2 + \cfrac{1}{1 + \cfrac{1}{2 + \cfrac{1}{1 + \cfrac{1}{1 + \cfrac{1}{4 + \ddots}}}}},$$

and since this continued fraction is both regular and infinite, he concluded that e is irrational.

It is usually pointed out that Euler did not make any explicit reference to what this infinity implies for e, that is, its non-rationality,

---

[26][Euler, 1744/1985, p. 311]. In the original: "Antequam ad alias fractiones continuas, in quibus denominatores progressiones arithmeticas constituunt, summandas progrediamur; quantitates quasdam transcendentes enoluamus, quae in fractiones continuas conuersaedent denominatores in progressione arithmetica progredientes, quo ex his via euadant planior eiusmodi fractiones continuas summandi. Hoc igitur logarithmis aliisque expressionibus transcendentibus tentans deprehendi in eiusmodi fractiones continuas deduci, si numerus cuius logarithmus hyperbolicus est vnitas, eiusque potestates quaeque considerentur." (at p. 120).

[27][Euler, 1744/1985, p. 311].

as in [Cretney, 2014, p. 147], where the author suggests that Euler might "have thought the result obvious or unimportant and did not wish to labour the point". But I think that Euler made its irrationality sufficiently clear throughout this paper. In this work. He was, in fact, considering from the start e as a non-rational quantity: "...let us analyze certain transcendental quantities ...", he write. And about these kinds of quantities (non-finitely-expressible inconmensurable quantities), as well as the irrational ones (finitely-expressible inconmensurable quantities), Euler had already written that they "will go across to a continued fraction running to Infinity". As a result, he was only at this point in the paper demonstrating to the reader, the property which he had already ascribed to e.[28]

However, the number e was a newcomer into the land of mathematics at that time, not acquiring the importance it currently has until the 19th century. The most widely-known special irrational was the number π, so the major breakthrough would really come from J. H. Lambert with the proof of its irrationality.[29]

Lambert in [Lambert, 1766/1770, part V], entitled *Vorläufige Kenntnisse für die, so die Quadratur und Rektifikation des Cirkuls suchen*,[30] commented that, as far as he knew, the question around the irrationality of the ratio between the diameter and the circumference of the circle is not yet closed, something that, according to him, could lead anyone to waste their time by searching for prospective fractions defining π. This is not a petty appreciation since around that time various authors showed up claiming to have squared the circle through the "exact" ratio-

---

[28]Euler's proof is not self-contained, as the readers can corroborate by themselves by going through the original paper, and even with the valuable support of [Cretney, 2014]) —who contends that Euler's main motivations for continued fractions came from his interest in solving the Riccati differential equation— there still seem to be some obscure areas.

[29]For an annotated Spanish translation of the two works of Lambert's mentioned below, see [Dorrego López and Fuentes Guillén, 2021] (a slightly modified English version of this book is in production by Springer. See [Dorrego López and Fuentes Guillén, 2023]). Among the most relevant English works on Lambert's contributions to this subject, I recommend the reader to go to [Baltus, 2003], [Berggren et al., 1997, pp. 205–374], [Serfati, 1992, pp. 56–83], [Serfati, 2018, pp. 179–184], and [Petrie, 2009].

[30]Preliminary Knowledge for Those Seeking the Quadrature and Rectification of the Circle.

Figure 2: Euler and Lambert, the driving forces towards the understanding of special irrationals in the 18th century (image: Euler, Leonhard - 1780 - Austria - Public Domain. `https://www.europeana.eu/item/92062/BibliographicResource_1000126145198s` and `http://ark.bnf.fr/ark:/12148/cb419202991`).

nal values for π. In any case, this treatise was not intended to rigorously dispel the doubts around this problem, but rather to make a gentle, non-academic introduction to the circle-squaring problem, recalling why it would be a waste of time pursuing this millennia-old problem by means of rational values for π.

It is later, in his [Lambert, 1761/1768], that Lambert did indeed provide a rigorous account for the irrationality of π (besides expanding on Euler's result by proving that all rational powers of e are irrational quantities).[31] Although Lambert's demonstration is difficult to follow, it is possible to grasp the general idea because the irrationality of π follows as an immediate consequence from a more general result: if v is a rational arc, its tangent $\tan v$ has to be an irrational quantity. Therefore, since $\tan \pi = 0$ is a rational quantity, π is irrational.

---

[31]I will be referring to this work as *Mémoire*.

Lambert used the continued fractions machinery to demonstrate this result. Upon applying Euclid's algorithm to the quotient $\tan v = \frac{\sin v}{\cos v}$, and obtaining its continued fraction:[32]

$$\tan v = \cfrac{1}{1 : v - \cfrac{1}{3 : v - \cfrac{1}{5 : v - \cfrac{1}{7 : v - \cfrac{1}{9 : v - \cdots}}}}},$$

Lambert proceeded over approximately thirty pages until achieving his goal.[33]

4 Legendre's Note VI in his *Élements de Géometrie* (1794)

Following Lambert's work on π, a new contribution in this vein appears in the work of Adrien-Marie Legendre (1752–1833), an important and influential mathematician in Paris.[34] In his *Éléments de géométrie*, first published in 1794, he took over Lambert's *Mémoire* and took the next natural step towards the elucidation of the nature of π by proving that π^2 cannot be a rational number. Furthermore when he presented his opinion about π after the proof, he prominently drew on what would become

---

[32]It is not possible to get into the details of Lambert's paper in this chapter due to its being too long. In any case, let us just say to provide some tip about Lambert's procedure, that he undertakes a formal division between the sine and the cosine by drawing on their expansions in infinite series.

[33]Lambert recognized in his *Vorläufige Kenntnisse* that his motivation for getting on with searching for continued fractions came from Euler's *Introductio in analysin infinitorum*, 1748 (Introduction to Analysis of the Infinite). As for Euler's *De Fractionibus Continuis Dissertatio*, Lambert made no reference to it, so it is likely that he was unaware of it (Lambert appears to have no problem with citing the authors he has drawn upon). Certainly, they maintained a correspondence from 1758 until 1771 (16 letters), but there is apparently no trace about irrational-related subjects, these letters being concerned with physics (see [Bopp, 1924]).

[34]For a biographical exposition of Legendre, see [Beaumont, 1861].

the modern theoretical framework for classifying irrationals, which was actually first introduced by Lambert in his *Mémoire*:[35] algebraic as representative of the property of being the root of an equation rather than the property of being expressible by a finite combination of algebraic operations:

> It is probable that the number π is not even included among algebraic irrationals, that is to say, it cannot be the root of an algebraic equation with a finite number of terms with rational coeficients; but it seems so difficult to prove rigorously this proposition.[36]

This is probably the first clearly modern expression in the classical period of what an algebraic number is. But in spite of the high influence of Legendre's book, it did not produce a quick change in the terminology or in the use of the aforementioned modern theoretical framework; rather, mathematicians kept using the property of being expressible as a yardstick to classify irrationals.

Legendre's *Éléments* had a clear educational objective, and it was for this reason that the author tackled the problem of π. Lambert's demonstration being difficult to follow, he decided to search for another way to prove its irrationality. In this context, the first edition is more insightful than others because Legendre made explicit the reason why he gives a different demonstration:

> We already know one proof of this proposition that has been given by Lambert in the Memoirs of Berlin, year 1761; but, as this proof is long and difficult to follow, we have tried to shorten and simplify it.[37]

The problem, he claimed, was not that Lambert's proof is not rigorous enough, as later authors would claim, but rather that the proof is difficult;[38] that is why the first edition[39] is more interesting than the

---

[35]Concretely in [Lambert, 1761/1768, p. 320].

[36][Legendre, 1794, pp. 303–304].

[37][Legendre, 1794, p. 296].

[38]See [Dorrego López and Fuentes Guillén, 2021, chapter 4].

[39]Also the second edition. I could not consult the third one, but from the fourth edition onwards, the reference to Lambert already appears in a footnote.

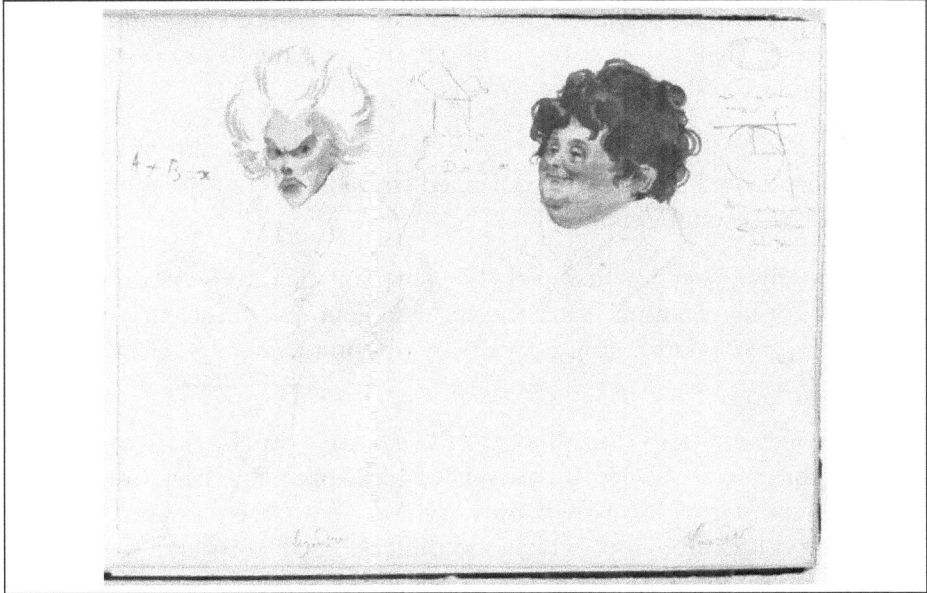

Figure 3: Adrien-Marie Legendre (left) along with Fourier (right). This caricature is the only known portrait of Legendre. Image from: Boilly, Julien-Leopold. (1820). *Album de 73 Portraits-Charge Aquarelle's des Membres de l'Institute* (watercolor portrait 29). Biliotheque de l'Institut de France (`https://commons.wikimedia.org/wiki/File:Legendre_and_Fourier_(1820).jpg`).

others in which the reference to Lambert is reduced to a footnote at the end of the proof: "This proposition was first demostrated by Lambert, in the Memoirs of Berlin, *anno* 1761". The overall idea of Legendre's proof can be easily grasped by means of the following outline:

- Legendre introduced an infinite series represented by $\varphi(z)$ as a departure point.

- From this series he obtained an expression in continued fraction for $\frac{a}{z} \cdot \frac{\varphi(z+1)}{\varphi(z)}$ that he calls $\psi(z)$.

- By comparing both expressions and putting $\frac{1}{2}$ as a concrete value for z, Legendre arrived at the continued fraction for the tangent.

- Then, Legendre proved two lemmas —the second lemma being weaker than the first one— about the irrationality of certain continued fractions.

- The milestone is the theorem in which he proves that the tangent is irrational because its continued fraction meets the requirements of the second lemma.

- Finally, Legendre advances the viewpoint that π is not a root of an algebraic equation using the term "algebraic" in a clearly modern way, and takes a step forward by proving that $\pi^2 \notin \mathbb{Q}$ (again, by means of the second lemma).

The relevance and the influence of Legendre's proof justify to include in this chapter an in-depth analysis of his Note. Furthermore, it will allow the reader to form their own opinion about the elegance or rigor of his procedure, both of which are difficult to denote and specify, as well as gain good knowledge about one proof of the irrationality of π that is considered a reference.[40]

4.1 The continued fraction for the tangent

Instead of borrowing the continued fraction for the tangent directly from Lambert and thereafter providing a simplified version of its irrationality for rational arcs, Legendre preferred to start from scratch and follow his own strategy to obtain that expression.

The point of departure was to consider the following infinite series:

$$\varphi(z) = 1 + \frac{a}{z} + \frac{1}{2} \cdot \frac{a^2}{z \cdot (z+1)} + \frac{1}{2 \cdot 3} \cdot \frac{a^3}{z \cdot (z+1) \cdot (z+2)} + \cdots,$$

that, upon some algebraic workout, yields a recurrent equality:

$$\varphi(z) - \varphi(z+1) = \frac{a}{z \cdot (z+1)} \cdot \varphi(z+2) \cdot$$

---

[40]For the sake of clarity, we will not be following Legendre's notation point by point, changing it slightly when necessary. Some additions to his exposition will be included in the spirit of making this proof fully self-contained.

This identity will be the base on which to obtain the sought-after continued fraction because now, after dividing it by $\varphi(z+1)$ and putting for the sake of brevity:

$$\psi(z) = \frac{a}{z} \cdot \frac{\varphi(z+1)}{\varphi(z)}, \tag{1}$$

Legendre achieved, a few simplifications later, an expression of $\psi(z)$ by means of a continued fraction:[41]

$$\psi(z) = \cfrac{a}{z + \cfrac{a}{z+1+\cfrac{a}{z+2+\cdots}}} .$$

But now, due to the fact that $\psi(z)$ —a continued fraction— is a function of $\varphi(z)$ —an infinite series—, it is possible to rewrite the equality (1)

---

[41]The process (not fully explained in the original) is as follows. Putting:

$$\psi(z) = \frac{a}{z} \cdot \frac{\varphi(z+1)}{\varphi(z)} ,$$

we obtain:

$$\frac{1}{\psi(z)} = \frac{z}{a} \cdot \frac{\varphi(z)}{\varphi(z+1)} \qquad \Longrightarrow \qquad \frac{a}{z \cdot \psi(z)} = \frac{\varphi(z)}{\varphi(z+1)} , \tag{2}$$

and from this last expression:

$$\frac{\varphi(z+2)}{\varphi(z+1)} = \frac{(z+1) \cdot \psi(z+1)}{a} . \tag{3}$$

Now, as Legendre said, dividing this identity by $\varphi(z+1)$ and using (2) and (3), we get:

$$\frac{\varphi(z)}{\varphi(z+1)} - 1 = \frac{a \cdot \varphi(z+2)}{z \cdot (z-1) \cdot \varphi(z+1)} \Longleftrightarrow$$

$$\Longleftrightarrow \frac{a}{z \cdot \psi(z)} - 1 = \frac{a \cdot (z+1) \cdot \psi(z+1)}{z \cdot (z+1) \cdot a} = \frac{\psi(z+1)}{z}$$

$$\Longleftrightarrow \frac{a}{z \cdot \psi(z)} = 1 + \frac{\psi(z+1)}{z} \Longleftrightarrow \frac{z \cdot \psi(z)}{a} = \frac{1}{1 + \frac{\psi(z+1)}{z}}$$

$$\Longleftrightarrow \psi(z) = \frac{a}{z + \psi(z+1)} .$$

Now we thus directly arrive at the continued fraction.

throughout both infinite processes:

$$\cfrac{a}{z + \cfrac{a}{z+1+\cfrac{a}{z+2+\cdots}}} = \frac{a}{z} \cdot \frac{1 + \frac{a}{z+1} + \frac{1}{2} \cdot \frac{a^2}{(z+1)\cdot(z+2)} + \cdots}{1 + \frac{a}{z} + \frac{1}{2} \cdot \frac{a^2}{z\cdot(z+1)} + \cdots}. \tag{4}$$

At this point, the continued fraction for the tangent will emerge from comparing these two expressions when $z = \frac{1}{2}$. On the one hand, it is obvious that the left side of the equality becomes:

$$\cfrac{2a}{1 + \cfrac{4a}{3 + \cfrac{4a}{5+\cdots}}}.$$

As for the right side, it becomes:

$$2a \cdot \frac{1 + \frac{4a}{3!} + \frac{16a^2}{5!} + \frac{64a^3}{7!} + \cdots}{1 + \frac{4a}{4!} + \frac{16a^2}{4!} + \frac{64a^3}{6!} + \cdots}.$$

But what we really have in both numerator and denominator after this substitution are exponential funtions. Bearing in mind that:

$$\frac{e^x - e^{-x}}{e^x + e^{-x}} = \frac{x + \frac{x^3}{3!} + \frac{x^5}{5!} + \frac{x^7}{7!} + \cdots}{1 + \frac{x^2}{2!} + \frac{x^4}{4!} + \frac{x^6}{6!} + \cdots},$$

it is easy, after some calculations, to bring the exponentials to light:

$$\sqrt{a} \cdot (2\sqrt{a}) \cdot \frac{1 + \frac{4a}{3!} + \frac{16a^2}{5!} + \frac{64a^3}{7!} + \cdots}{1 + \frac{4a}{2} + \frac{16a^2}{4!} + \frac{64a^3}{6!} + \cdots} =$$

$$= \sqrt{a} \cdot \frac{(2\sqrt{a})\cdot 1 + (2\sqrt{a})\cdot\frac{4a}{3!} + \cdots}{1 + \frac{4a}{2!} + \frac{16a^2}{4!} + \cdots}$$

$$= \sqrt{a} \cdot \frac{2\sqrt{a} + \frac{(2\sqrt{a})^3}{3!} + \frac{(2\sqrt{a})^5}{5!} + \cdots}{1 + \frac{(2\sqrt{a})^2}{2!} + \frac{(2\sqrt{a})^4}{4!} + \cdots}$$

$$= \sqrt{a} \cdot \frac{e^{2\sqrt{a}} - e^{-2\sqrt{a}}}{e^{\sqrt{a}} + e^{-2\sqrt{a}}}.$$

Thus, taking up once again the equality (4) (interchanging both sides each other), this means that:

$$\frac{e^{2\sqrt{a}} - e^{-2\sqrt{a}}}{e^{2\sqrt{a}} + e^{-2\sqrt{a}}} \cdot 2\sqrt{a} = \cfrac{4a}{1 + \cfrac{4a}{3 + \cfrac{4a}{5 + \cdots}}}.$$

The hidden nature of this equality showed up after Legendre set $4a = -x^2$,[42] because in this case, taking into account Euler's formulas and that $2\sqrt{a} = ix$, we have:[43]

---

[42]Legendre said that from this equality "two principal formulas are derived according as a is positive or negative" ([Legendre, 1794, p. 298]). For the second case he set $4a = -x^2$, from which he will obtain the continued fraction for the tangent. But for the first one, in the case that $4a = x^2$, Legendre just showed the expression without further comments:

$$\frac{e^x - e^{-x}}{e^x + e^{-x}} = \cfrac{x}{1 + \cfrac{x^2}{3 + \cfrac{x^2}{5 + \cdots}}}.$$

This is in reality the hyperbolic tangent expressed in a continued fraction, a continued fraction that will also fall under the conditions of the theoretic results he will use in this note in order to prove the irrationality of the (circular) tangent for rational arcs. The irrationality of the hyperbolic tangent had already been proven by Lambert in his *Mémoire*, who, along with Vincenzo Riccati, made one of the first uses of the hyperbolic functions as such (see [Barnett, 2004] for details).

[43]Let us bear in mind that by using these formulas:

$$\begin{aligned} e^{xi} + e^{-xi} &= 2\cos x \\ e^{xi} - e^{-xi} &= i \cdot 2\sin x, \end{aligned}$$

we obtain:

$$i \cdot \tan x = i \cdot \frac{2\sin x}{2\cos x} = \frac{e^{xi} - e^{-xi}}{e^{xi} + e^{-xi}}.$$

$$\frac{e^{2\sqrt{a}} - e^{-2\sqrt{a}}}{e^{2\sqrt{a}} + e^{-2\sqrt{a}}} \cdot 2\sqrt{a} = \cfrac{4a}{1 + \cfrac{4a}{3 + \cfrac{4a}{5 + \cdots}}} \Longleftrightarrow$$

$$\Longleftrightarrow \frac{e^{ix} - e^{-ix}}{e^{ix} + e^{-ix}} \cdot ix = \cfrac{-x^2}{1 + \cfrac{-x^2}{3 + \cfrac{-x^2}{5 + \cdots}}}$$

$$\Longleftrightarrow \frac{e^{ix} - e^{-ix}}{e^{ix} + e^{-ix}} \cdot i = \cfrac{-x}{1 + \cfrac{-x^2}{3 + \cfrac{-x^2}{5 + \cdots}}}$$

$$\Longleftrightarrow \tan x = -\frac{e^{ix} - e^{-ix}}{e^{ix} + e^{-ix}} \cdot i = \cfrac{x}{1 + \cfrac{-x^2}{3 + \cfrac{-x^2}{5 + \cdots}}},$$

and finally:

$$\tan x = \cfrac{x}{1 - \cfrac{x^2}{3 - \cfrac{x^2}{5 - \cfrac{x^2}{7 - \cdots}}}} \cdot$$

Irrationality criterion for certain continued fractions

Once Legendre obtained the continued fraction for the tangent, the strategy he followed was centered on proving two lemmas about the irrationality of certain continued fractions. The second of these two lemmas will be, in reality, a weaker version of the first one. Let us let Legendre speak for himself:

LEMMA I. *Let,*

$$\cfrac{m}{n + \cfrac{m'}{n' + \cfrac{m''}{n'' + etc.}}},$$

be a continued fraction prolonged to infinity in which all the numbers m, n, m', n', etc. are positive or negative integers. If the component fractions $\frac{m}{n}$, $\frac{m'}{n'}$, $\frac{m''}{n''}$, etc, are all supposed to be less than unity, I say that the total value of the continued fraction will be an irrational number.[44]

The proof of this lemma is comprised of two parts: "Firstly I say that this value will be less than unity",[45] excepting in the case that this continued fraction has the form:[46]

$$\cfrac{m}{m + 1 - \cfrac{m'}{m' + 1 - \cfrac{m''}{m'' + 1 - \cdot \cdot \cdot}}}.$$

This holds since all the convergents are less than unity, and it is easy to convince oneself of this by just analysing some of them, for instance, the second convergent:

$$\frac{p_2}{q_2} = \cfrac{m}{n + \cfrac{m'}{n'}}.$$

Firstly, we have by hypothesis that $|\frac{m}{n}| < 1$. This means that $|m| < |n| = n$, and due to that fact that both are integers, $|m| \le n - 1$. On

---

[44][Legendre, 1794, pp. 298–299]. Reading the proof, it is clear that he is speaking about absolute values, that is to say $|\frac{m^{(i)}}{n^{(i)}}| < 1$ (see also [Baltus, 2003, p. 8]. On the other hand, Legendre did not tackle the convergence of the continued fraction.

[45][Legendre, 1794, p. 299].

[46]The value of this continued fraction is exactly the unity because if we calculate the convergents $\frac{p_n}{q_n}$ we would obtain:

the other hand, $n - 1 < |n + \frac{m'}{n'}|$ because $|\frac{m'}{n'}| < 1$. Putting these two facts together we have that $|m| < n-1 < |n+\frac{m'}{n'}|$, and therefore $|\frac{p_2}{q_2}| < 1$.

As Legendre wrote:

> By continuing the same reasoning, we shall see that, whatever number of terms of the proposed continued fraction be calculated, the value resulting from them is less than unity; hence the total value of this continued fraction prolonged to infinity is also less than unity.[47]

$$\frac{p_1}{q_1} = \frac{m}{m+1}$$

$$\frac{p_2}{q_2} = \frac{mm' + m}{(mm' + m) + 1}$$

$$\frac{p_3}{q_3} = \frac{mm'm'' + mm' + m}{(mm'm'' + mm' + m) + 1}$$

$$\cdots$$

Therefore $q_n = p_n + 1$, and hence:

$$\frac{p_n}{q_n} = \frac{p_n}{p_n + 1} = \frac{1}{1 + \dfrac{1}{p_n}} \xrightarrow[n \to \infty]{} 1,$$

because if $n^{(i)} > 0$, then $m^{(i)} > 0$ and in this case $p_n \xrightarrow[n \to \infty]{} \infty$. Take into account that Legendre let $n, n', n'', \cdots$, be positive integers, but as he himself commented this does not really affect the generality of the theorem. This is because if any of them is negative, let us put $n' = -j$, then:

$$\cfrac{m}{n + \cfrac{m'}{-j + \cfrac{m''}{n'' + \cdots}}} = \cfrac{m}{n + \cfrac{-m'}{j + \cfrac{-m''}{n'' + \cfrac{m'''}{n''' + \cdots}}}} = \cfrac{m}{n + \cfrac{m'}{n' + \cfrac{m''}{n'' + \cdots}}}.$$

[47][Legendre, 1794, pp. 299–300].

This being proven, the second part is carried out by supposing the rationality of the continued fraction:

$$\frac{B}{A} = \cfrac{m}{n + \cfrac{m'}{n' + \cfrac{m''}{n'' + \cdots}}}.$$

Legendre let C, D, E, etc. be indeterminate quantities such that:

$$\frac{C}{B} = \cfrac{m'}{n' + \cfrac{m''}{n'' + \cfrac{m'''}{n''' + \cdots}}},$$

$$\frac{D}{C} = \cfrac{m''}{n'' + \cfrac{m'''}{n''' + \cfrac{m^{IV}}{n^{IV} + \cdots}}},$$

and so on to infinity, and what he is going to prove is that these indetermite quantities are actually integer quantities that in absolute value form a decreasing succession:[48]

- They are integers because A, B, and all $m^{(i)}$ and $n^{(i)}$ are also integers, so:

$$\frac{B}{A} = \cfrac{m}{n + \cfrac{C}{B}} \quad \Longrightarrow \quad C = mA - nB \in \mathbb{Z},$$

$$\frac{C}{B} = \cfrac{m'}{n' + \cfrac{D}{C}} \quad \Longrightarrow \quad D = m'B - n'C \in \mathbb{Z},$$

---

[48]Legendre is using here Fermat's method of infinite descent.

$$\frac{D}{C} = \frac{m''}{n'' + \dfrac{E}{D}} \implies E = m''C - n''D \in \mathbb{Z},$$

etc.

- Now, we have $0 < \cdots < |D| < |C| < |B| < |A|$ because $|B| < |A|$, $|C| < |B|$, $|D| < |C|$, etc, since, according to the first part of this demonstration, $|\frac{B}{A}|$, $|\frac{C}{B}|$, $|\frac{D}{C}|$, etc are less than unity.[49]

This lemma is the basic pillar on which all this note is set up. The second lemma loosens the conditions over $|\frac{m^i}{n^i}|$ by letting them be less than unity only "after a certain interval" onwards:

LEMMA II. *The same things being established, if the component fractions $\frac{m}{n}$, $\frac{m'}{n'}$, $\frac{m''}{n''}$, etc. are of whatever magnitude at the beginning of the series, but after a certain interval they are constantly less than unity, I say that the proposed continued fraction, always supposing that it tends to infinity, will have an irrational value.*[50]

This also guarantees the irrationality of the continued fraction, because if we establish that it is from $\frac{m'''}{n'''}$ onwards that all the fractions $\frac{m'''}{n'''}$, $\frac{m^{IV}}{n^{IV}}$, $\frac{m^V}{n^V}$, etc are less than unity:

then by lemma 1, the continued fraction

$$\frac{m'''}{n''' + \dfrac{m^{IV}}{n^{IV} + \dfrac{m^V}{n^V + \text{etc.}}}}$$

will have an irrational value. Call this value ω, and the continued fraction will become

$$\frac{m}{n + \dfrac{m'}{n' + \dfrac{m''}{n'' + \omega.}}}$$

---

[49]Legendre pointed out that:

besides no one of the numbers A, B, C, D, E, etc. can be zero, since the proposed continued fraction goes on to infinity...

[50][Legendre, 1794, p. 302].

But if we successively put

$$\frac{m''}{n'' + \omega} = \omega', \quad \frac{m'}{n' + \omega'} = \omega'', \quad \frac{m}{n + \omega''} = \omega''',$$

it is clear that, ω being irrational, all the quantities ω', ω'', ω''', must be so likewise. But ω''' the last of these is equal to the proposed continued fraction; hence the value of this fraction is irrational.[51]

4.2 Irrationality of the tangent for rational arcs. Implications

At this point, the irrationality of the tangent for rational arcs turns out to be an immediate consequence. Anyway, Legendre labeled it as a theorem due to its relevance:

THEOREM.

If an arc is commensurable with the radius, its tangent will be incommensurable with the same radius.[52]

If we let the arc be a rational quantity $x = \frac{m}{n}$, the continued fraction for the tangent becomes:

$$\tan \frac{m}{n} = \cfrac{m}{n - \cfrac{m^2}{3n - \cfrac{m^2}{5n - \cfrac{m^2}{7n - \cdots}}}}.$$

Now it is obvious that this continued fraction falls under the conditions of the second lemma because:

whilts the numerator m^2 continues of the same magnitude, the component fractions will be or will soon become less than unity; hence the value of $\tan \frac{m}{n}$ is irrational; therefore; *if the arc is commensurable with the radius, its tangent will be incommensurable.*

---

[51][Legendre, 1794, p. 302].
[52][Legendre, 1794, p. 303].

From this there follows as an immediate consequence the proposition which is the object of this note. Let π be the semicircumference of which the radius is 1; if π were rational, the arc $\frac{\pi}{4}$ would also be, and by consequence its tangent should be irrational: but the tangent of the arc $\frac{\pi}{4}$ is known on the contrary to be equal to the radius 1; therefore π cannot be rational. Hence *the ratio of the circumference to the diameter is an irrational number.*[53]

But Legendre did not stop here, going beyond his analysis of π by using his method to prove the irrationality of π^2. This is also easily derived from the above-established second lemma, due to the fact that if we rewrite $0 = \tan \pi$ in terms of continued fractions:[54]

$$0 = 3 - \cfrac{\pi^2}{5 - \cfrac{\pi^2}{7 - \cfrac{\pi^2}{9 - \text{etc.}}}},$$

and let $\pi^2 = \frac{m}{n}$ be a rational quantity, it is plain that the following equality cannot be held:

$$3 = \cfrac{m}{5n - \cfrac{m}{7 - \cfrac{m}{9n - \cfrac{m}{11 - \cdots}}}},$$

because the resultant continued fraction also falls under that lemma.

---

[53][Legendre, 1794, p. 303]. Just here is where the reference to Lambert from the fourth edition onwards appeared as a footnote (I repeat that I could not consult the third edition).

[54]This is because when a continued fraction is inverted "it goes up a position". In this case:

$$0 = \cfrac{\pi}{1 - \cfrac{\pi^2}{3 - \cfrac{\pi^2}{5 - \cdots}}} \implies \frac{1}{0} = 1 - \cfrac{\pi^2}{3 - \cfrac{\pi^2}{5 - \cfrac{\pi^2}{7 - \cdots}}} \implies 0 = \frac{1}{\frac{1}{0}} = 3 - \cfrac{\pi^2}{5 - \cfrac{\pi^2}{7 - \cfrac{\pi^2}{9 - \cdots}}}.$$

4.3 Reflections and conjectures

Throughout his note, Legendre follows his own strategy in order to establish his results, although he is conspicuously drawing on Lambert's. Moreover, as we have already mentioned, it was Lambert who for the first time established the modern distinction between incommensurable quantities in terms of algebraic (roots of equations) vs non-algebraic (non-roots of equations), whereas the tradition had differentiated between incommensurable quantities by taking expressibility, instead of the property of being root of an equation, as a yardstick. Just before extending Lambert's proof, Legendre, relying heavily on Lambert's approach, established what is probably the first clear definition of what an algebraic quantity is, conjecturing the transcendence (in the modern sense of the term) of π:

> It is probable that the number π is not even included among algebraic irrationals, that is to say it cannot be the root of an algebraic equation with a finite number of terms with rational coeficients; but it seems so difficult to prove rigorously this proposition; we can only show that the square of π is also an irrational number.[55]

5 Concluding remarks

There is no doubt that Euler and Lambert were two of the most brilliant scholars of the 18th century. They were giants. There is little we can add to the existing knowledge of Euler, he being widely known as the most brilliant and prolific mathematician of the 18th century (and maybe beyond). But we should maybe emphasize that, even if Lambert is not as well known as Euler or other 18th-century scholars, he was considered as a leading researcher by the likes of such giant figures as Euler himself or Lagrange, or further as a genius, as in the case of Kant (to name a few).[56]

---

[55][Legendre, 1794, pp. 303–304]. Lambert had already conjectured this property before in his *Mémoire* (see [Lambert, 1761/1768, p. 321].

[56]Lagrange in letter to D'Alembert dated on 15 July 1769 wrote: "M. Lambert, on whom you wish to know my opinion, is undoubtedly one of the best individuals

There is no doubt, on the other hand, that both of these two brilliant men were at the center of academic life in Europe. Euler was at that time a member of the Russian Academy of Sciences in St. Petersburg. Interestingly, the Academy served not just as an Academy, but also as both a university and secondary-school center. Opened in 1725, its *Commentarii academiae scientiarum Petropolitanae*,[57] where the Academy's members published his contributions, rapidly acquired a wide reputation: "I cannot tell you often enough how eagerly people everywhere ask for the St Petersburg Mémoires $\cdots$".[58] The same could be said in the case of Lambert, who was a member of the Berlin Academy of Science. From 1746, it was one of Europe's leading centers, with members including Maupertuis, Lagrange, and Euler himself until his definite departure to St. Petersburg (1766).[59] But neither the giantness of the actors nor the relevance of these institutions were enough to disseminate these results, which would today be looked on as so remarkable. This was not only because investigations of this type were, at this time, far from being mainstream, but also because the tool used to investigate these highly theoretical issues at the time was in its first stages of development.[60]

As for the connections between Lamberts and Legendre's proofs, the differences and similarities are quite clear. For instance, Lambert's proof is noticeably longer, in part because he spent ten pages rigurously proving the convergence of the continued fraction of the tangent (something that Legendre did not), a step worth mentioning if we take into account the standards of rigour at that time.[61] Legendre's, on the other hand,

---

of our Academy; a hard worker who practically holds up alone our Class of Physics [...]" (in [Lalanne, 1882, p. 141]).

[57] Memoirs of the Academy of Sciences in St. Petersburg.

[58] Daniel Bernoulli in a letter to Euler dated on 7 December 1734 (quoted in [Lipski, 1953, p. 353].

[59] As for the Berlin Academy of Science see [Aarsleff, 1989].

[60] Expositions about continued fractions included in 18th-century sources leave out their applicability to studying irrational quantities, apart from their power to approximate them. The *Encyclopédie* (1777) may serve as a model: the reader can see the article dedicated to continued fractions (where, by the way, Lambert is not mentioned) in [Brezinski, 1991, pp. 324–326].

[61] Note that it cannot be said that 18th-century mathematicians were not worried about foundational questions. We can actually enumerate many cases: the status of negative and complex numbers, Euclid's fifth-postulate problem, the existence of infinitely small quantities, or the debates around divergent series. [Schubring, 2005]

is more general: he established some irrationality-related results that will be applied to the concrete case of the tangent, and Lambert dealt at all times just with the tangent. Interestingly, there is another difference that epitomizes the two main methodological approaches: while Lambert did not hide his cards and showed from the very outset his ideas —he started out with a continued fraction that he deduced in a constructive and intuitive way, a clearly heuristic approach— Legendre brought an infinite series into being apparently out of nothing in order to deduce the required expression for the tangent, and used general results for his benefit. In other words, while Legendre appeared to "efface his tracks like a fox" in some steps (in the Gaussian style), Lambert acted with an overwhelming level of detail, which is likely why Glaisher wrote:

> Although Legendre's method is quite as rigorous as that on which it is founded, still, on the whole, the demonstration of Lambert seems to afford a more striking and convincing proof of the truth of the proposition.[62]

Although we also run into the other face of the coin:

> his investigation, however, is given in such detail, and so many properties of continued fractions, now well known, are proved, that it is not very easy to follow his reasoning, which extends over more that thirty pages.[63]

This is probably one of the reasons why some authors in the past considered Legendre's proof to be more elegant. This is the case, for instance, of François Joseph Servois (1768–1847) in his article devoted to Lambert, in which he found a place to include as one of Lambert's relevant results "the celebrated proof" of the irrationality of π. In any case, he commented that, to be honest, it should be said that this proof has

---

has drawn attention to this issue: " It is a widespread opinion to think of eighteenth-century mathematics as unconcerned with the foundations and as interested only in the further development of analysis. As we have seen with the concepts of negative numbers as well as with the infinitely small quantities, the mathematicians were, in contrast, very anxious to clarify basic concepts." ([Schubring, 2005, p. 285].

[62][Glaisher, 1871, p. 12].

[63][Glaisher, 1871, p. 12].

gained in elegance and, above all, simplicity, after passing through the hands of Legendre.[64]

In any case, there is an important similarity apart from the fact that they both use the same method (continued fractions): the rigorous style that turns their proofs into undeniable justifications for the age and, in essence, for ours.

References

[Aarsleff, 1989] H. Aarsleff. The Berlin Academy under Frederick the Great. *History of the Human Sciences*, 2(2): 193–206, 1989.

[Baltus, 2003] C. Baltus. Continued Fractions and the first proofs that pi is irrational. *Communications in the Analytic Theory of Continued Fractions*, Vol. XI, 5–24, 2003.

[Barnett, 2004] J. H. Barnett. Enter, stage center: the early drama of the hyperbolic functions. *Mathematics magazine,* 77(1), 15–30, 2004.

[Beaumont, 1861] Elie de Beaumont. Éloge historique de Adrien-Marie Legendre. *Mémoires de l'Académie des Sciences*, 32, 1861.

[Berggren et al., 1997] L. Berggren, J. Borwein, and P. Borwein. (eds.). *Pi: A source book.* Springer, New York, 1997.

[Bopp, 1924] K. Bopp. Leonhard Eulers und Heinrich Lamberts Briefwechsel aus den manuskripten herausgegeben. *Aus den Abhandlungen der Preussischen Akademie der Wissenschaften, Phys.-Math. Klasse*, 2, Berlin, 1924.

[Bos, 2001] H. J. M Bos. *Redefining Geometrical Exactness: Descartes' Transformation of the Early Modern Concept of Construction.* Springer, New York, 2001.

[Brezinski, 1991] C. Brezinski. *History of Continued Fractions and Padé Approximants.* Berlin: Springer, 1991.

---

[64] [Servois, 1819, p. 48]. Michel Serfati in his insightful study of the development of irrational and transcendental quantities in the 18th and 19th centuries, summarizes —from our point of view quite rightly— Legendre's method and style in comparison with other 18th- and 19th-centuty authors as follows (see [Serfati, 1992, p. 84]):

> Compared with Lambert and Euler, but also with Hermite and the mathematicians of the nineteenth century, Legendre's mathematical method and style of exposition are very modern, almost contemporary, that is to say at once elegant, and on the other hand, abrupt and obscure.

[Cataldi, 1613] P. Cataldi. *Trattato del modo brevissimo di trouare la radice quadra delli numeri, et regole da approssimarsi di coninuo al vero nelle radice de'numeri non quadrati, con le cause, et inventioni loro. Et anco il modo di pigliarne la radice cuba, applicando il tutto alle operationi militari et altre.* Cochi, Bartolomeo, 1613.

[Cretney, 2014] R. Cretney. The origins of Euler's early work on continued fractions. *Historia Mathematica,* 41(2): 139–156, 2014.

[Crippa, 2014] D. Crippa. *Impossibility results: from geometry to analysis.* Doctorat en Epistémologie et Histoire des Sciences, Universite Paris Diderot, 2014.

[Crippa, 2019] D. Crippa. *The Impossibility of Squaring the Circle in the 17th Century: A Debate among Gregory, Huygens and Leibniz.* Birkhäuser, Basel, 2019.

[Dorrego López, 2021] E. Dorrego López. *Irrationality and transcendence in the 18th and 19th centuries. A contextualized study of J. H. Lambert's Mémoire (1761/1768).* PhD Dissertation. Universidad de Sevilla, 2021.

[Dorrego López and Fuentes Guillén, 2021] E. Dorrego López, E. Fuentes Guillén (prefacio de José Ferreirós). *Dilucidando π. Irracionalidad, trascendencia y cuadratura del círculo en Johann Heinrich Lambert (1728–1777).* College Publications, Cuadernos de Lógica, Epistemología y Lenguaje, 2021.

[Dorrego López and Fuentes Guillén, 2023] E. Dorrego López, E. Fuentes Guillén (foreword by José Ferreirós). *Irrationality, Transcendence and the Circle-Squaring Problem. An Annotated Translation of J. H. Lambert's* Vorläufige Kenntnisse *and* Mémoire. Springer, Logic, Epistemology, and the Unity of Science, 2023.

[Ebbinghaus et al., 1988] H. D. Ebbinghaus, H. Hermes, F. Hirzebruch, M. Koecher, K. Mainzer, J. Neukirch, A. Prestel, and R. Remmert. *Zahlen (2nd edition), Grundwissen Mathematik,* Springer-Verlag, 1988. References to the English translation: *Numbers.* Springer, New York, 1995.

[Euler, 1744/1985] L. Euler. De fractionibus continuis dissertatio. *Commentarii academiae scientiarum Petropolitanae.* Vol. 9: 98–137, 1744. References to the English translation: B. Wyman. An Essay on Continued Fractions. *Math. Systems Theory,* Num. 18, 295–328, 1985.

[Favaro, 1875] A. Favaro. *Notizie storiche sulle frazioni continue: dal secolo decimoterzo al decimosettimo.* Tip. delle scienze matematiche e fisiche, 1875.

[Fraser, 1990] C. Fraser. Lagrange's Analytical Mathematics, Its Cartesian Origins and Reception in Comte's Positive Philosophy. *Studies in History and Philosophy of Science,* 21(2): 243–256, 1990.

[Glaisher, 1871] J. W. L. Glaisher. On Lambert's Proof of the Irrationality of

π, and on the Irrationality of certain other Quantities. *Report of the British Association for the Advancement of Science, 41st. Meeting, Edinburgh,* 12–16, 1871.

[Heath, 1908] T. Heath. *The thirteen books of Euclid's Elements. Translated from the text of Heiberg, Vol II, Books III–IX.* Cambridge, The University Press, 1908.

[Knobloch, 2006] E. Knobloch. Beyond Cartesian limits: Leibniz's passage from algebraic to "transcendental" mathematics. *Historia Mathematica,* 33: 113–131, 2006.

[Lalanne, 1882] L. Lalanne. *Oeuvres de Lagrange: t.13 Correspondance inédite de Lagrange et d'Alembert, publiée d'après les manuscrits autographes et annotée par L. Lalanne.* Paris: Gauthier-Villars, 1882.

[Lambert, 1761/1768] J. H. Lambert. Mémoires sur quelques propriétés remarquables des quantités transcendantes, circulaires et logarithmiques. *Mémoires de l'Académie royale des sciences de Berlin,* 265–322, 1761/1768.

[Lambert, 1766/1770] J. H. Lambert (1766/1770). *Beyträge zum Gebrauche der Mathematik und deren Anwendung II.* Berlin, Band II in zwei Theilen, 1766/1770.

[Legendre, 1794] A. M. Legendre. *Éléments de géométrie, avec des notes, (1st edition).* F. Didot (Paris), 1794.

[Lipski, 1953] A. Lipski. The Foundation of the Russian Academy of Sciences. *Isis,* 44(4), 349–354, 1953.

[Petrie, 2009] B. J. Petrie. Euler, Lambert, and the Irrationality of e and π. *Proceedings of the Canadian Society for History and Philosophy of Mathematics,* 22, 104–119, 2009.

[Petrie, 2012] B. J., Petrie. Leonhard Euler's Use and Understanding of Mathematical Transcendence. *Historia Mathematica,* 39(3), 280–291, 2012.

[Schubring, 2005] G. Schubring. *Conflicts between Generalization, Rigor, and Intuition. Number Concepts Underlying the Development of Analysis in 17–19th Century. France and Germany.* New York: Springer, 2005.

[Sereda, 2017] K. Sereda. *Leibniz on the Concept, Ontology, and Epistemology of Number.* PhD dissertation, UC San Diego, 2017.

[Serfati, 1992] M. Serfati. *Quadrature du cercle, fractions continues et autres contes. Sur l'histoire des nombres irrationnels et transcendants aux XVIII et XIX siècles.* Brochure A.P.M.E.P., n° 86, 1992.

[Serfati, 2018] M. Serfati. *Leibniz and the invention of mathematical transcendence.* Franz Steiner Verlag, Stuttgart, 2018.

[Servois, 1819] F. J. Servois. Lambert (Jean-Henri). In: *Biographie universelle ancienne et moderne,* 23, 46–51, 1819.

[Smith, 1958] D. E. Smith. *History of Mathematics Vol. 2*. Dover edition, 1958.

[Stedall, 2000] J. A. Stedall. Catching Proteus: the collaborations of Wallis and Brounker. I. Squaring the circle. *Notes and records of the Royal Society of London*, 54(3), 293–316, 2000.

[Wallis, 1656/2004] J. Wallis. *Arithmetica infinitorum*. Oxford, 1656. References to the English translation: Jacqueline A. Stedall. *The Arithmetic of Infinitesimals: John Wallis 1656*. Translated from Latin to English with an introduction by Jacqueline A. Stedall. Springer, New York, 2004.

Wendlingen: A Scientist in the Eighteenth Century Spanish Court

Joaquim Berenguer Clarià
Universitat Politècnica de Catalunya, Spain
jberenguer90@gmail.com

Abstract

In the mid-eighteenth century, the Spanish Bourbon Monarchy wanted to link science with the task of renovation of the institutions based on the well-being and the progress of the country. The still powerful Society of Jesus was one of the sectors at the service of the Crown in the process of "modernization" where science and technology were identified with material progress of the society. Around 1750 the Society of Jesus sent Johannes Wendlingen (1715-1790), who was a Jesuit mathematician and teacher in Prague, to the Imperial College in Madrid. Wendlingen became professor of mathematics in this college, a Cosmographer of the Indies and tutor of the princes. He was charged by the Spanish Crown to establish an astronomical observatory in the Imperial College that would have connections with the network of Europeans observatories. The aim of our paper is, on the one hand, to show Wendlingen as an active agent of applying mathematics to the technological development of the country, that is, as a technician supporting economic and military interests of the Spanish Crown. On the other hand, we want to emphasize Wendlingens role as the one in charge of writing a complete course on mathematics, as a textbook for teaching in the Imperial College. The part of this treatise that is dedicated to differential calculus deserves special attention since it shows us Wendlingen as one of the introducers of infinitesimal analysis in Spain, a very scarcely known field at that time in this country.

Project PID2020-113702RB-I00: Matemáticas, Ingeniería y Patrimonio: Nuevos Retos y Prácticas, XVI-XIX' of the Ministerio de Ciencia e Innovación of Spain. Centre de Recerca per a la Història de la Tècnica

1 Introduction

As many historians[1] have already analysed, eighteenth-century Spain was still a society based on a rural and feudal economy. However, under the Bourbon monarchy a certain reform movement took place that had already begun in the late 17th century, based on the development of commercial trade and new manufacturing industries that gave rise to an increasing number of bourgeoisie and craftsmen.

Science was an essential factor for promoting this renewal movement that in fact served to consolidate the Monarchy. Nevertheless, a stable body was needed in order to receive this new knowledge. And the Society of Jesus became one of the institutions capable of implementing and guiding the new science. It is well known that the Society of Jesus, which for a long time had operated as a social and political pressure group, used its influence to train and educate members of the most powerful classes, and that with the emergence in the mid-18th century of new social sectors allied to more "liberal" pressure groups, the different enlightened monarchies in Europe sought to rid themselves of the influence of the Jesuits. Even so, the Jesuits' task of disseminating knowledge, particularly scientific knowledge, cannot be overlooked. According to [Romano, 1999], the Jesuits appear as actors in scientific life, where their ability through the college to organise the circulation of knowledge throughout academic networks provided them with a privileged position as transmitters of this knowledge. According to [Navarro Brotóns, 2001], the role played by the Jesuits in the dissemination of science during the 18th century became stronger throughout Europe thanks to the widespread networking of the members of the Society. Indeed, for much of the seventeenth and eighteenth centuries, the chair of mathematics at the Imperial College of Madrid was the vehicle for the transmission of European science in Spain.[2] Be that as it may, the conservatism of the Society of Jesus and its close links with political and economic were no an impediment to it becoming an institution that transmitted the new scientific disciplines in 18th century Spain. Both the institution and its members adopted the discourse of "modernity", understood as science

---

[1][De la Torre Ruiz, 2013]; [López Piñero, 1969];[Sellés *et al.*, 1988]; [Valverde Pérez, 2007].

[2][Berenguer, 2016].

at the service of utility, while at the same time strengthening the new enlightened monarchy.

The Jesuit Johann Wendlingen (1715–1790) is a good example of this highly complex relationship between science and political power at the Spanish Court, under the kingdoms of Felipe VI and Carlos III.

The aim of our paper is, firstly, to show that Wendlingen, astronomer of the Court, Cosmographer of the Indies and tutor of the princes, was an active agent of mathematics applied to the technological development of the country, that is, as a technician working on the behalf of the economic and military interests of the Spanish Crown.

Secondly, we focus on the manuscripts written by Wendlingen and his Mathematics textbooks published during his time in Spain. We especially stress Wendlingen's concern to reinforce the practical aspect of his texts. In order to understand his role in the introduction of the new calculus into eighteenth century Spain, it is particularly interesting to analyse his manuscripts on Differential and Integral Calculus.

Thirdly, analysis of the very severe criticisms that Wendlingen received from his contemporaries will help to qualify the general conclusions about the role of this scientist, who for more than fifteen years exercised a considerable influence in the Spanish Court.

Our article consists of six parts, the first of which is devoted to a brief description of Wendlingen's biography, in which his stay in Madrid is of particular significance. The second part addresses Wendlingen's tasks as a scientist at the Court of Madrid. The third and fourth parts deal with the mathematical treatises written by Wendlingen during his stay in Madrid, both those published and those preserved in manuscript form. The fifth part refers to the criticisms levelled at Wendlingen's work during his stay in Spain, and finally, in the sixth part we draw some conclusions about Wendlingen's role in Spain.

2 Wendlingen: a Jesuit scientist sent to the Court of Madrid

Johann Wendlingen was born in Prague. He joined Society of Jesus in 1734 after four years spent as a teacher of Humanities at the Clementinum Jesuit College of Prague. After becoming a doctor of Philosophy, he moved in 1750 to the Spanish Court as a Mathematics teacher at the Imperial College of Madrid, in accordance with the expansionist policy of the Jesuits in Europe. He was subsequently appointed the Chief Cosmographer of the Indies and under the patronage of Ensenada[3] he was able to set up a new astronomical observatory at the Imperial College with the instruments acquired by Jorge Juan[4] in London. In 1759 he was named mathematics tutor of the Prince of Asturias — the future king Carlos IV — as well as the other princes. According to [Navarro Brotóns, 2001], Wendlingen was sent to Madrid on the recommendation of the king's confessor Francisco Rávago, and thus joined the Spanish Court through this very Royal intervention. It is necessary to point out that he continued to receive a pension as Royal tutor from the Spanish Court after his expulsion from Spain.

In 1761, Christian Rieger (1714-1780) arrived from Vienna in order to replace Wendlingen as the main teacher of mathematics and as a Cosmographer of the Indies. Wendlingen however continued to impart his classes at the Imperial College until the Jesuits were expelled from Spain in 1767, when he returned to Prague and remained there for two years in the position of director of the Museum of Mathematics at the Clementinum College of Prague. At the end of his life he moved to Liběšice where he died in August, 1790, at the age of 75.[5]

---

[3]Zenón de Somodevilla y Bengochea, Marquis de la Ensenada (1707–1781) was a State Councillor during the kingdoms of Felipe V, Fernando VI and Carlos III. He was Minister of Finance, the Navy, War and the Indies from 1743 to 1754. During this period, while Ferdinand VI was reigning, he established the bases for the creation of a powerful Spanish navy and promoted the diffusion of science in the army.

[4]Jorge Juan (1713–1773), a Spanish naval engineer and scientist, collaborated with Ensenada to modernise the Spanish navy. He promoted the dissemination of science in the navy and is the author of several scientific texts.

[5]"Liběšice" is the town name given in [O'Neill et al., 2001], which corresponds to a present-day town in the Czech Republic. In [Pelzel, 1786] and [Backer et al., 1960] the name given is "Libeschitz" and in [Poggendorff, 1863], "Liboschütz".

3   Wendlingen's commitments at the Court of Madrid

The main position assigned to Wendlingen was that of Chief Cosmographer of the Indies, an office created by the Spanish Court a couple of centuries earlier. After the discovery of the New World, the Spanish Crown, eager to consolidate its position in the transoceanic maritime routes, encouraged the study and development of cartography and cosmography. The Crown first created the *Casa de Contratación* in Seville (1503) and subsequently set up new institutions such as the Council of the Indies (1524) and established various posts such as that of Chief Cosmographer (1563). These institutions and posts promoted both the economic exploitation of the new domains and the development of science and techniques linked to navigation and cosmography. The Chief Cosmographer of the Indies, in particular, had to report to the Council on everything concerning the geographical discoveries of the New World. Among other responsibilities, this position involved controlling the prospecting and exploitation of mines as well as cartographic tasks.

From 1628 onwards, the Council of the Indies delegated the teaching of mathematics and cosmography to the Society of Jesus at the Imperial College. The Royal Council proposed the appointment of the Chief Cosmographer and first professor of Mathematics, positions that had to be held by the same person. who had to be an outstanding member of the Society of Jesus.[6] Thus, upon his arrival in Madrid, Wendlingen assumed both the post of Chief Cosmographer of the Indies and that of first professor of mathematics at the Imperial College.

The second task that the Spanish Crown commissioned Wendlingen to undertake was to set up an astronomical observatory in Madrid.

It was essential for the Spanish Monarchy to achieve a well-founded production of scientific knowledge comparable with that which already existed in Europe. It was therefore essential to build this new astronomical observatory in Madrid with the instruments obtained through Jorge Juan. Wendlingen was the person responsible for implementing this project in collaboration with Solferino,[7] and, later with Antonio Luís

---

[6] See in [De la Torre Ruiz, 2013] and [Sánchez Martínez, 2010].

[7] Duque de Solferino: Francisco de Gonzaga y Pico de la Mirandola, Member of

Real y Lombardón, Esteban Bramieri, Antonio Zacagnini and Miguel Marín[8] ([Valverde Pérez, 2007, p. 196]). Some months later, a new astronomical observatory was opened in Cádiz and managed by Jorge Juan and Louis Godin.[9] And in this way an incipient network directed by Joseph Nicolas Delisle (1688–1768)[10] in Paris was established. The correspondence between Wendlingen and Deslile shows Wendlingen's involvement in some astronomical research projects between 1750 and 1761:[11]

Felipe V's Chamber and Royal Steward to the Queen.

[8]Antonio Luís Real y Lombardón, who became a general lieutenant in the Navy, was a disciple of Wendlingen.

Esteban Bramieri (1720–1794) was a Jesuit who became famous for his participation in the measurements of the geographical boundaries between Paraguay and Brazil. From 1757 he was a professor at the Seminary of Nobles in Madrid.

Antonio Zacagnini (1723–1803), a Jesuit, was professor of physics at the Seminary of Nobles in Madrid from 1755 to 1767. Miguel Marín (1725-?), a Jesuit, was a teacher from 1754 at the Seminary of Nobles in Calatayud.

[9]Louis Godin (1704-1760) was director of the Royal Academy of Marine Guards in Cadiz and is the author of the treatise *Compendium of Mathematics*.

[10]Delisle was a French astronomer who organized a worldwide study of a transit of Venus (1761), the first such systematic study to be made.

[11][Valverde Pérez, 2007, p. 218]. The author has obtained the data from the *Correspondance de M. Delisle*. Archive National de Paris. *Marine*, 2JJ/66 i 2JJ/68.

Year	
1750	4th, 6th and 20th of December: Transit of the first Satellite of Jupiter 13th of December: Eclipse of the Moon (T. XI n° 78b)
1751	6th of February: Transit of the first Satellite of Jupiter 1st and 14th of March (T. XI n° 95)
1753	6th of May: Transit of Mercury over the Sun 11th, 16th and 24th of May: Transit of the Satellites of Jupiter 21st of October: Eclipse of the Sun (T. XII n° 146)
1754	3rd, 10th and 12th of May: Emersions and immersions of the three Satellites of Jupiter 4th, 13th and 20th of June: Emersions of the first and the second Satellites of Jupiter Observations to determine the maximum declination of the Ecliptic (T. XIII n^c 94)
1755	Wendlingen is working on the twin meridian lines of Escorial. He is collecting data to determine the obliquity of the Ecliptic (T. XIII n°179)
1756	He is working in the palace for building the new meridian line (T. XIII n° 179)
1757	7th of July: Eclipse of the Moon (T. XIV n° 9a, b, c)
1758	24th of January: Eclipse of the Moon (T. XIV n° 15a, b)
1761	Observations of the transit of Venus (T. XV n° 24)

In 1754 Wendlingen was commissioned by the King Carlos III to build two identical meridian lines in the Palace-Monastery of Escorial. After building these two meridian lines,[12] Wendlingen was charged with building another meridian line in the park of El Retiro in Madrid, whose construction is described in the work *Explicación y uso de la Meridiana*, 1756 (see figure 1).

A reference to an observation of the Eclipse of the Moon in 1757 is preserved: "Observatio Eclipsis Lunaris facta a Matriti a P. Johanne Wendlingen, Societate Jesu in Regali Observatorio Collegii Imperiales ejusdem Societatis", which appeared in *Philosophical transactions* (fig.

---

[12]A meridian line is a gnomic instrument by which it is possible to show the solar midday. The meridian lines were built for solving the problem of the connection of the spring equinox with Easter, which was a concern of the Catholic Church.

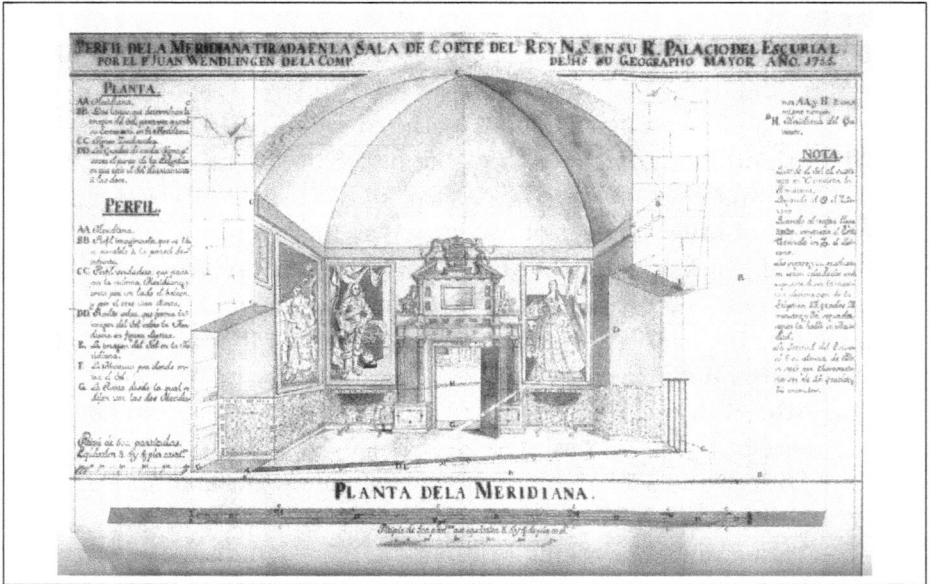

Figure 1: Drawing of the meridian installed in the Court Room of the Escorial, by Wendlingen in 1755. *Escorial Palace, Madrid, Spain. Side view, royal cabinet bedroom.* Madrid Spain, 1755. Photograph. `https://www.loc.gov/item/ade2002000001/`.

2).

Wendlingen was thoroughly devoted to the astronomical project managed by Delisle, as well as to the European astronomical endeavour in the early 1740s, the purpose of which was mainly to solve the following three problems: 1) To improve the solar theory: to determine the obliquity of the elliptic 2) To correct the mistakes generated by refraction and the instrumental failures, and 3) To correct the position of the stars, keeping the aberration in mind. Wendlingen sought to secure the collaboration of his master Stepling[13] in Prague, as well as attempting to establish a plan for extending the astronomical observations to America using his position as Cosmographer of the Indies.

---

[13] Joseph Stepling (1716–1778) was director and founder of the Prague Astronomical Observatory and also director of the mathematical library in Prague. He was Wendlingen's teacher while Wendlingen was in Prague before his move to Spain.

LXXXVIII. *Obſervatio Eclipſis Lunaris faƀa*
Matriti *a P*ª. Joanne Wendlingen, *Socie-*
tatis Jeſu, *in Regali Obſervatorio Collegii*
Imperialis ejuſdem Societatis, Die 30 Julii
1757.

Quælibet obſervatio bis inſtituta fuit, ſemel inter-
jeƀo oculum inter lentemque ocularem vitri clari,
cærulei, plani, ac bene terſi, fragmento. Hæ ob-
ſervationes notantur hac voce cerul. *Teleſcopium,*
quo uſus ſum, eſt Gregorianum *trium pedum* Angli-
canorum, *omnino præclarum.*

Communicated by Matthew Maty, *M. D. F. R. S.*

[Read April 20, 1758.]

IMMERSIONES.		Tempus verum			Differentia
		h	′	″	
PRincipium eclipſis, *clar.* -		9	47	34	
Mare Humorum, *clar.* -		—	52	47	″
Grimaldus - - - - {*cær.*		—	54	28	— 59
{*clar.*		—	55	27	
Bullialdus - - - - {*cær.*		10	1	21	— 13
{—		10	1	34	
Keplerus - - - {—		—	9	35	— 10
{—		—	9	45	
Copernicus - - {—		—	16	15	— 12
{—		—	16	28	
Heraclides - - - {—		—	18	14	— 10
{—		—	18	24	
				Manilius	

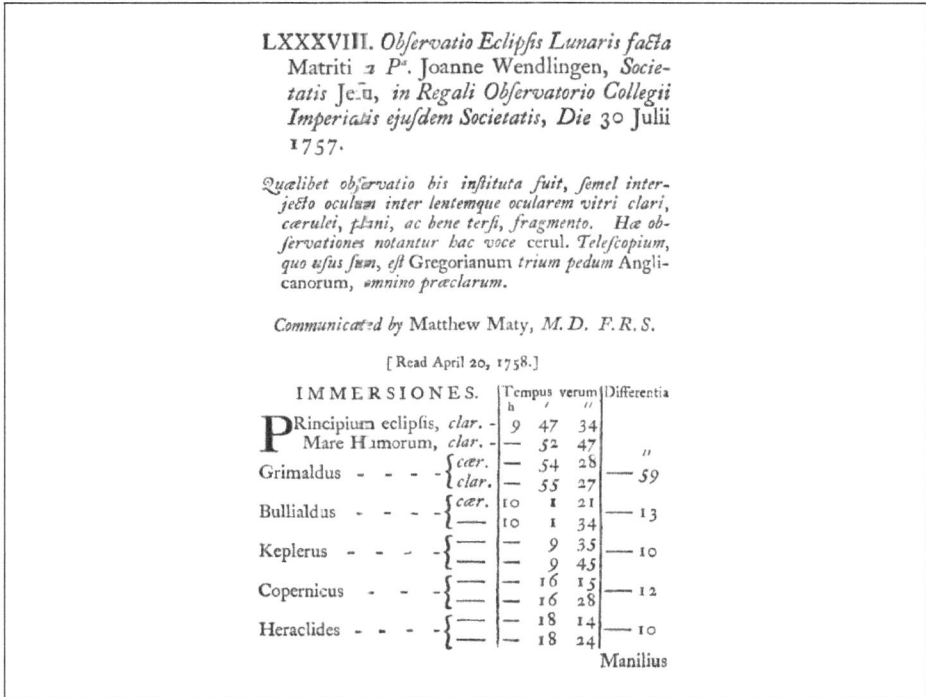

Figure 2: *Philosophical Transactions of the Royal Society.* London. Volume 50, p. 640-645. https://archive.org/details/philtrans09135565.

However, the reason for Wendlingen's move to the Royal Court in Madrid was not confined to the creation of an astronomical observatory, but also to a further concern of the monarchy, which was the establishment of a mathematical museum. A Mathematics Museum already existed at Wendlingen's former place of study, Clementinum College, where a collection of astronomical instruments were on display. It was probably for this reason, that Wendlingen's arrival at Madrid was associated with the desire of the Court to create a similar Museum. According to Valverde ([2007, p. 79]), there was a real fascination at the Court of Madrid for machines and the launch of the new observatory was linked to a request from the nobles for a Mathematics Museum.

When Rieger replaced Wendlingen in 1761, the project for the Mathematics Museum had been quite forgotten because of the need for more practical premises for the teaching of experimental Physics.

4 The publication of a textbook: the *Elementos de la mathematica*

While he was in Spain, and probably by order of the Crown, he planned to publish a mathematics work in several volumes, four of which were published between 1753 and 1756. They comprise the *Elementos de la mathematica*; the first volume is devoted to arithmetic, the second to geometry, the third to logarithms, plane and spherical trigonometry, while the fourth contains tables of sinus, tangents and logarithms. According to Wendlingen himself, this text was inspired by the treatise *Elementa Matheseos Universæ* (1713–1715) by Christian Wolff (1679–1754), but with a more practical orientation. This is understandable if we bear in mind that one of the pressing needs of the Spanish Monarchy was the training of technicians to drive the economic and social modernisation that the country required, such as making an updated map of the country. One of the purposes of the book, therefore, was for teaching geographers and topographers.

4.1 Wendlingen's volume on Arithmetic

At the beginning of his first volume on arithmetic, Wendlingen includes a prologue in which he first of all states the social and economic usefulness of applied mathematics:

> In the interests of the public good, who can deny that the Mathematical Faculties are the only ones whose knowledge, in practical terms, fills with happiness the Kingdoms and Provinces wherever they flourish? [...]They provide farmers with the measurement of their fields; they determine their capacities for the seed; they teach them how to build machines, whether for vital daily use or for the construction of great and luxurious buildings, as well for the building City walls [...] when put into practice, the knowledge of these

Faculties provides the quickest and safest way for making Monarchies both rich and happy [...][14]

He then explains that it is necessary to include in Mathematics a range of branches focused on application to the social and economical development of the country. Specifically the branches that Wendlingen considers in this prologue are as follows: Arithmetic, Plane Geometry, Stereometry, Plane and Spherical Trigonometry, Algebra, Mechanics, Hydrostatics, Aerometric, Hydraulic, Optics, Catoptrics, Perspective, Astronomy, Geography, Art of Navigation, Chronology, Gnomonic, Pyrotechnics, Military and Civil Architecture.

It seems that this list corresponds to the subjects he wanted to cover in a course, although they are not exactly those that appear in his treatise *Elementos de la mathematica* or in the manuscripts that have been preserved by this author. In any event, he ends the prologue with a reference to Wolff:

The method will be the same as that of Wolff, from whom I depart very little by always trying to combine the practical to the speculative, so that the pleasure of practice sweetens the work of study, and enlivens the application.[15]

This is followed by a section entitled "Prolegómenos" which is partly based on the section "De Methodo Mathematica. Brevis Commentario" that Wolff also includes in the first volume of his *Elementa Matheseos Universæ*. This is a text in which the general method used in mathematics is explained and the concepts of axioms, postulates, propositions, scholia, etc. are clarified.

---

[14]"Si se atiende al bien público, quien negará a las Facultades Matemáticas, a ser las únicas, cuya noticia, reducida a la práctica, llena de felicidades los Reinos y Provincias en donde florecen?[..] Miden a los labradores los campos; determinan sus capacidades para la semilla; les enseña a hacer máquinas, tanto para el uso cotidiano muy necesarias, como a propósito a construir edificios magníficos y suntuosos; como también a murar las Ciudades, [...] la noticia de estas Facultades puesta en práctica es el medio más breve y seguro de enriquecer y felicitar las Monarquías [...]" ([Wendlingen, 1753–1756, "Tomo I", "prologo"]).

[15]"El modo será el mismo que el de Wolffio, de quien discordaré en poco, procurando siempre unir lo práctico a lo especulativo, para que lo gustoso de la práctica endulce el trabajo del estudio, y avive la aplicación". ([Wendlingen, 1753–1756, "Tomo I", "prologo"]).

Wendlingen's volume on Arithmetic is inspired by '"Elementa Arithmeticæ", included in Wolff's *Elementa Matheseos Universæ*, and although Wendlingen does not clearly distinguish sections in his treatise, the similarities between the two texts may be observed in the following table:

Wolff's "Elementa Arithmeticæ"	Wendlingen's Volume I on Arithmetic
Caput Primum. De Principiis Arithmeticæ	Principles of arithmetic
Caput II De speciebus Arithmeticæ in numeris integris	Arithmetic operations: adding, subtracting, multiplying, partitioning
Caput III De Ratione ac proportione quantitatum	Ratios and proportions
Caput IV De speciebus Arithmeticæ in numeris fractis	Fractions
Caput V De Potentiis numerorum, Genesi præsertim ac Analysi numerorum Quadratorum et Cubicorum	Powers and roots
Caput VI De Regulis Proportionum	Rules of proportion
Caput VII De Quantitatibus Æquidifferentibus	
Caput VIII De Logarithmis	
Caput IX De Fractionibus Decimalibus	Decimal arithmetic
Caput X De Fractionibus Sexagesimalibus	Sexagesimal arithmetic[16]

---

[16]"Principios de la aritmética; Operaciones aritméticas: sumar, restar, multiplicar, partir; Razones y proporciones; Quebrados; Potencias y raíces; Reglas de la proporción (regla de tres, de ganancias, de aligación); Aritmética decimal; Aritmética sexagesimal" ([Wendlingen, 1753–1756, "Tomo I. Aritmetica"]).

4.2 Wendlingen's interest in the practical application of Geometry

The second volume is entitled "Elements of Practical Geometry" ("Elementos de la Geometria Practica") and consist of three parts:

Part I: Elements of Plane Geometry ("Parte I: Elementos de la Geometria Llana")

Part II: On the principal properties of Planes, as regards their Sections and Positions ("Parte II: De las principales propiedades de los Planos, en cuanto a sus Secciones y Posiciones")

Part III: Elements of Solids, or of Stereometry ("Parte III: Elementos de los Solidos, o de la Estereometria")

The first part is the longest and is divided into four sections. In the first section Wendlingen introduces the basic definitions of geometric elements, several theorems and some practical problems of measuring distances and constructing figures. The second section includes some problems of areas of surfaces. In the third section there are problems of calculation of angles and distances on the terrain, and the fourth section is entitled "Ichnography", defined by Wendlingen himself when he writes that "to describe a figure ichnographically is to reduce the large to small with due proportion".[17]

As in the volume on Arithmetic, Wendlingen largely follows Wolff's text "Elementa Geometriæ" included in the *Elementa Matheseos Universæ*. It is not a mere copy, since Wendlingen does not reproduce the structure of Wolff's text, but shortens its content and, above all, emphazises the practical dimension of the problems.

Let us consider some problems from Wendlingen's volume on Geometry, contrasting them with the corresponding ones by Wolff, where the Bohemian mathematician's interest in the practical application of geometry is obvious:

---

[17]"[...] describir 'ichnographicamente' una figura es reducir lo grande a pequeño con la proporción debida." ([Wendlingen, 1753–1756, "Tomo II. Geometria", §302]).

Wolff's "Elementa Geometriæ"[18]	Wendlingen's Practical Geometry[19]
Problema IX	Problem XIV
194. Metiri distantiam duorum locorum $A\&B$ ex eodem tertio C accessorum.	148. Measure the distance between two points AB, which can be approached from the arbitrary point C by the Geometer.

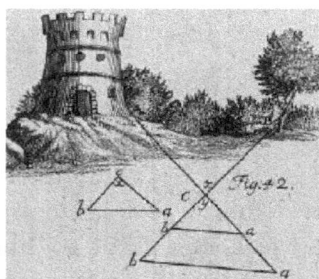

Figure 3. ([Wolff, 1713-1715, §194].)

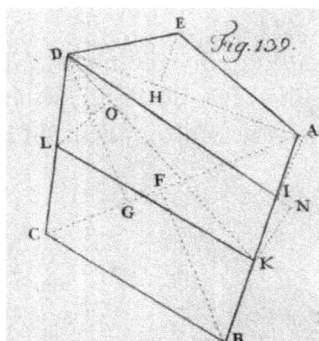

Figure 6. ([Wendlingen, 1753–1756, "Tomo II", §148])

Problem LXVI	Problem LXV
441. Figuram rectilineam quamcunque $ABCDE$ in partes æquales dividere.	257. Break the surface of any rectilinear figure into three or more parts in order that the Geometer moves away from acute angles, that is to say, no part has excessively acute angles.

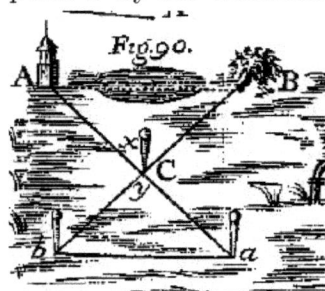

Figure 4. (Wolff 1713–1715, "Elementa Geometriæ", §441)

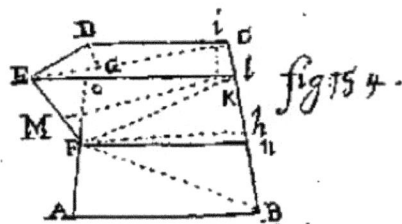

Figure 7. ([Wendlingen, 1753–1756, "Tomo II", §257])

Scholium

258. We have posed this Problem here for those who are distributing fields, particularly in inheritances, because if we do not pay attention to the acute angles we may cause great harm to one or more inheritors.

Problema XXVIII	Problem LXXX
286. Altitudinem inaccessam AB metiri.	299. To find the height of a house, tower, etc. that is on a mountain, and the Geometer cannot approach either the mountain or the level of the building.

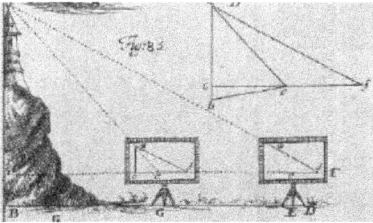

Figure 5. (Wolff 1713–1715, "Elementa Geometriæ", §286) Problem LXXX

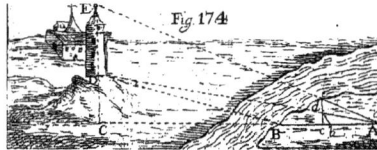

Figure 8. ([Wendlingen, 1753–1756, "Tomo II", §299])

---

[18]([Wolff, 1713-1715, "Elementa Geometriæ", §194, §441, §286]).

[19]"Problema XIV. 148. Medir la distancia de dos lugares AB, a los cuáles desde el punto arbitrario C se puede arrimar el Geómetra; Problema LXV. 257. Partir el área de cualquiera figura rectilinea en tres, o más partes iguales de modo que se aparte el Geómetra de los ángulos agudos: quiero decir que ninguna parte tenga ángulos demasiadamente agudos; Escolio. 258. *Se ha puesto este Problema aquí para que los que estan haciendo reparticiones de los campos, especialmente en casos de las herencias, porque no haciendo caso de los ángulos agudos, se puede perjudicar mucho al uno, o más de los herederos*; Problema LXXX. 299. Hallar la altura de una casa, torre, & c. que esté en una montaña, y el Geómetra no puede arrimarse, ni a la montaña, ni al piso de la fábrica". ([Wendlingen, 1753–1756, "Tomo II", §148, §257, §258, §299]).

It is worth pointing out the scholium that appears with problem LXV, where Wendlingen wishes to stress its usefulness when dealing with the distribution of land in the case of inheritance.

If we compare problem LXV with the corresponding Wolff problem LXVI, we see immediately that the statements are different. In Wolff's statement it is only a question of dividing a linear figure into equal parts without further conditions. In Wendlingen's case, it is a question of breaking a linear figure down into equal parts but in such a way that excessively acute angles are avoided.

Firstly, consider the resolution of the exercise according to Wolff:

First of all, the total area of the figure has to be calculated and divided by three. Wolff considers the first triangle AED (see figure 4) and calculates the difference between this area and the third part of the total area. This difference can be the area of a triangle AID, whose height will be the result of dividing the area by half the base AD. By drawing a parallel to AD at a distance equal to the calculated height, we obtain the intersection l of this parallel with the side AB. The area of the figure $AEDI$ will be one third of the total area. Next, a triangle IKD is constructed which is one sixth of the total area, using the same method as above, and then another triangle KLD is added which is also one sixth of the total area, so that the figure $KIDL$ will effectively be one third of the total area. Let us note that by considering two triangles whose area is one-sixth of the total area, instead of one triangle whose area is one-third allows Wolff to construct the quadrilateral $KIDL$ which will effectively have fewer acute angles. Thus, although he does not make it explicit, he also prevents the resulting parts from having angles that are too acute. Finally the remaining figure $LKBC$ will also have an area of one third of the total area.[20]

---

[20]"Resolutio

1. Quæratur area figuræ& dividatur in tot partes æquales, in quot figura dividi debet, ex. gr. In 3.

2. Area partis, in nostro casu tertiae, ulterius dividatur bifariam.

3. Area trianguli AED subtrahatur a parte tertia, & residuum dividatur per 1/2 AD; erit quotus altitudo trianguli AID priori AED addendum, ut AEDI fit pars tertia figuræ.

4. Quare intervallo hujus altitudinis ducatur parallela ipsi AD quæsecabit latus AB in I: quo puncto dato, rectam DI ducere licet, tertiam partem figuræAIDE abscindentem.

5. Pars tertia dimidia, sive sexta totius figuræ, dividatur per 1/2 DI quotus erit

Now let us see how this problem is solved by Wendlingen:

Let the figure $ABCDEF$ (see figure 9) be divided into three equal parts. The total area of the figure has to be calculated and this area divided by three. Furthermore, like Wolff, Wendlingen calculates the area of the triangle EDC and obtains the difference between this area and the third part of the total area calculated. This difference will be represented by a triangle of base EC. Moreover, as in Wolff's exercise, Wendlingen calculates the height iK of this triangle by dividing the area of the triangle by half of the base, i.e. half of EC. A line is then drawn parallel to EC at a distance equal to the height found. The point of intersection of this parallel with the side CB will be l. The area of the polygon $EDCl$ will be one third of the total area. Up to this point Wendlingen follows the same steps as Wolff, but from here on he then draws the line Fl and calculates the area of the triangle lFE. As he did before he calculates the difference between this area and the third part of the total area. He thus finds the triangle Fln that must be added (or subtracted) to make the area of the figure effectively one third of the total area. $ElnF$ will have as area this third part. This gives two thirds of the area, so $FnBA$ will also have one third of the area.[21]

---

altitudo trianguli IKD sextam figuræpartem constituentis.

6. Intervallo igitur hujus altitudinis agatur ipsi ID parallela, ut habeatur punctum K

7. Dividatur quoque dimidia pars tertia figuræper 1/2 KD, ut habeatur altitudo trianguli KLD sextæitidem parti figurææqualis.

8. Quare hujus intervallo denuo agatur ipsi KD parallela, ut punctum L determinetur ducaturque recta KL, quæpartem figurætertiam $KIDL$ resecabit.

9. Si figura in plures quam tres partes resolvenda; eodem modo ulterius procedentum". ([Wolff, 1713-1715, "Elementa Geometriæ", §441]).

[21]"Sea la figura $ABCDEF$ la que se ha de partir en tres partes iguales. I) Pártase toda la figura en cuatro triángulos, para hallar la capacidad de toda la figura. II) Pártase la suma de todos los triángulos por tres, y el coto dará la porción que toca a cada uno. III) Empiécese la repartición de la figura por el triángulo EDC, y véase cuanto le falta para ser la tercera parte, o que le sobra. IV) la falta v.g. que se ha hallado puede representar el área de un triángulo, cuya altura se halla partiendo el área por la mitad de la base; esto es, por la mitad de la EC, y sale la altura iK. V) Levántese esta en la EC perpendicularmente en donde quiera. VI) Tírese por el punto K una paralela Ml a la EC, para que se halle el punto l en donde la Fn corta a la CB. VII) Tírese la El, y será la $EDCl$ la tercera parte de toda la figura dada. VIII) Tírese la Fl y búsquese la capacidad del triágulo lFE, y se verá lo que le falta para ser la tercera parte de toda la figura. IX) El triángulo que le falta se hallará del

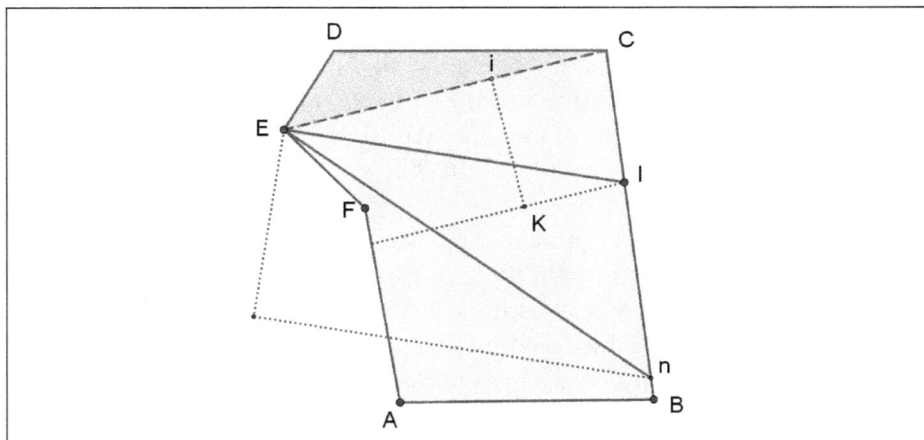

Figure 9: Author's own drawing on the basis of Wendlingen's original figure (Figure 7).

Therefore, the difference between Wolff's method and Wendlingen's method, starting from the first polygon whose area is the required area (in our case the third part) is based on the fact that the first method adds triangles whose area is the desired part of the total area, taking as a base the last side constructed from the previous polygonal figure. However, Wendlingen repeats the procedure followed in order to obtain the first polygon, making a new provisional triangle (in our case IFE) from a new vertex of the total figure, which in our case is the vertex F, and so on. The consequence of this procedure, which at first sight seems more complex, is that the parts into which the total figure is divided do not contain such acute angles, which was Wendlingen's aim. This is because he takes the vertices of the original total figure as the origin of the lines that will delimit each of the parts into which the total figure will be divided. Wendlingen's method is best adapted to the particular shape of the initial figure. Wendlingen is indeed thinking not only as a mathematician but also as a topographer.

It is worth considering the result we would obtain if the Wendlingen

mismo modo que se halló el de arriba. X) Luego tírese Fn y se hallarán las dos tercias partes. Lo que resta ya se ve que es una tercera parte. Si los triángulos son mayores se quitará de ellos la parte necesaria del mismo modo". ([Wendlingen, 1753–1756, "Tomo II", §257, §258]).

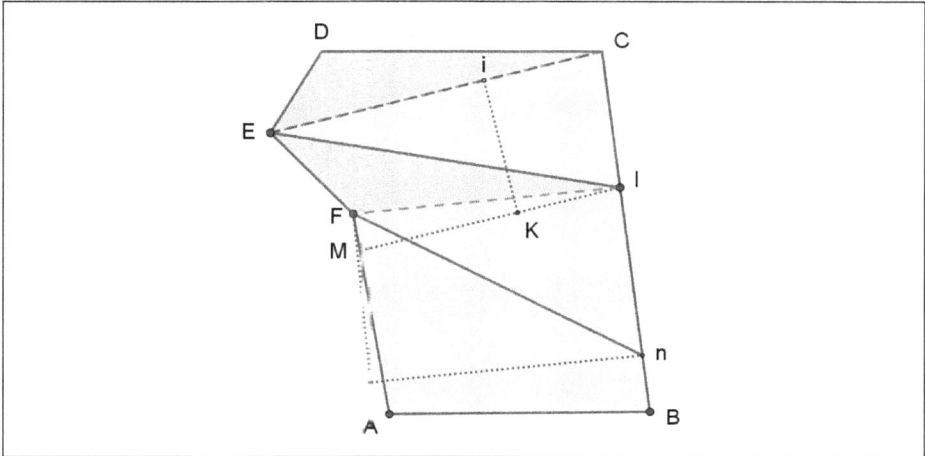

Figure 1C: Author's own drawing

figure were to be divided according to Wolff's method (see figure 10).

Wolff's method does not take into account the particular shape of the initial figure, and it is only possible to make the triangle Eln as a second part. In this case, the last part $EnBAF$ would have been significantly disadvantaged with respect to the first two parts.

4.3   Volums III and IV of *Elementos de la mathematica*

The third volume is entitled "Logarithmic, Plane Trigonometry, Spherical Sections, and Spherical Trigonometry" ("Logarithmica, Trigonometria Plana, Secciones de la Esphera, y Trigonometria Espherica") and consists of three parts, as follows: "Treatise I on Logarithmics" ("Tratado I de la Logarithmica"), "Part II on Plane Trigonometry" ("Parte II de la Trigonometria Llana") and "Part III on Spherical Trigonometry" ("Parte III de la Trigonometria Espherica"). It appears that Wendlingen does not follow Wolff's text so faithfully where trigonometry is concerned, especially spherical trigonometry, since in the *Elementa Matheseos Universae* there is no section dealing with this subject. In this volume, Wendlingen proposes trigonometric problems in which the questions refer to situations similar to those that military engineers might be confronted with:

Figure 11: ([Wendlingen, 1753–1756, "Tomo III", "Parte II; "De la Trigonometría Llana", §49])

Problem XII

49. The Engineer wants to build the wall B from A, without knowing either the distance AB whether the artillery will reach the objective. On completing his Geometrical Operations, he was unable to find more than the hypotenuse BC, with the angle BCA, even though he knows that BAC is a rectangle triangle. How will he find the requested distance BA?[22]

Problem XVII

61. A Coastguard sees Pirate A approaching Castle B. He wants to repel him with his artillery but does not know if he is within range. Question: How does he know this trigonometrically?[23]

---

[22]"Problema XII. 49. El Ingeniero quiere desde A batir la muralla B, sin saber la distancia AB, ni si la artillería hará el efecto. No pudo, habiendo hecho sus Operaciones Geométricas, hallar más que la hipotenusa BC, con el ángulo BCA, aun sabiendo que BAC es un triángulo rectángulo. Cómo hallará la distancia pedida BA?" ([Wendlingen, 1753–1756, "Tomo III", "Parte II", "De la Trigonometria Llana", §49]).

[23]"Problema XVII. 61. Un Guarda-costas ve que el Pirata A se arrima al Castillo B. Quiere apartarle con la artillería que tiene pero no sabe si alcanza. Pregúntase: Cómo lo sabrá trigonométricamente?" ([Wendlingen, 1753–1756, "Tomo III", "Parte II", "De la Trigonometria Llana", §61]).

Figure 12: ([Wendlingen, 1753–1756, "Tomo III", "Parte II; De la Trigonometría Llana", §61])

Volume IV contains a collection of logarithmic, sines and tangent tables which, as stated on the title page, are attributed to Wolff:

TABLES OF SINES AND TANGENTS

Both natural and artificial. On logarithms for natural numbers and of square and cubic numbers, starting from the unit to 1000.

Reduced to the present form by Christian Wolff.

In many parts amended and extended with an Instruction for easy handling, both in Arithmetical Accounts and in the Operations of the two Trigonometries.[24]

In fact, Volume IV reproduces Wolff's text *Tabulæ sinuum atque tangentium tam naturalium quam artificialium* (1728) in its entirety, adding

---

[24]"Tablas de los Senos y Tangentes

Así naturales como artificiales. De los logaritmos para los números naturales y de los números cuadrados y cúbicos, empezando desde la unidad hasta 1000.

Reducidas a la forma presente por Don Christiano Wolfio.

En muchas partes enmendadas y aumentadas con una Instrucción para el manejo fácil de ellas, así en Cuentas Aritméticas como en las Operaciones de las dos Trigonometrías".

([Wendlingen, 1753–1756, "Tomo IV", front page]).

several examples to the introductory part where Wendlingen explains how to use these tables.

5 Wendlingen's manuscripts

In the Royal Academy of the History of Madrid we find some volumes saved as manuscripts, which are the continuation of the other four volumes of *Elementos de la mathematica* already published by Wendlingen, and thus may assume that they were written between 1756 and 1761:

- Volume V. Algebra applied to Elementary Geometry ("Tomo V. Algebra aplicada a la Geometria Elemental")

- Volume VI. Algebra applied to Sublime Geometry ("Tomo VI. Algebra aplicada a la geometria sublime")

- Volume VII. Algebra applied to Arithmetic ("Tomo VII. Algebra aplicada a la Arithmetica")

- Volume VIII. Analysis of the infinities ("Tomo VIII. Analysis de los infinitos")

- Volume IX. Exponential Calculus, Differentio-differential Calculus and Arithmetic of the infinities ("Tomo IX. Calculo Exponencial, Diferencio-diferencial y Arithmetica de los infinitos")

These handbooks are in strict accordance with Wolff's text *Elementa Matheseos Universæ*, as Wendlingen himself recognizes:

> [...] I am not following my own whim but rather the famous Christian Wolff, who without doubt and in a methodical way is the Prince of all Mathematicians, which is why those who read my *Elementos* will not be surprised if in most part of them they find the same propositions, names and operations of Wolff, because I have merely translated them.[25]

---

[25] "[...] no sigo en esto mis antojos propios sino al célebre Christiano Wolffio, el cual sin controversia, mirando lo metódico, es el Príncipe de todos los Matemáticos, por lo cual no extrañen los que leyesen mis elementos, hallando en ellos, en la mayor parte, las mismas proposiciones, denominaciones y operaciones de Wolffio, los cuales he traducido solamente". ([Wendlingen, 1756–1761, "Tomo IX", Cortes 9/2812, f. 59]).

Indeed, if we consider Wendlingen's published *Elementos* together with his unpublished manuscripts, it can be seen that his aim was to publish a complete treatise on mathematics, taking Wolff's work as a guide, as can be seen in the following comparative table:

Wolf	Wendlingen
Elementa Matheseos Universæ	*Elementos de la mathematica*
Elementa Arithmeticæ	Volume I Arithmetic
Elementa Geometriæ	Volume II Geometry
Elementa TrigonometriæPlanæ	Volume III Logarithmic, Plane Trigonometry, Spherical Sections, and Spherical Trigonometry
Tabulæsinuum atque tangentium tam naturalium quam artificialium	Volume IV Table of sines, and tangents
Elementa Matheseos Universæ, **Elementa Analyseos Mathematicæ, Pars Prima, Sectio Secunda de Algebra**	
Caput Primum De Algebra ad Problemata arithmetica, eaque determinata, applicata Caput II De Algebra ad Problemata Arithmetica indeterminata applicata	Volume VII. Algebra applied to Arithmetic Section I: Definitions and Hypotheses necessary for understanding analytical operations Section II: On the application of Algebra to given Arithmetical Problems
Caput III De Algebra ad geometriam elementarem applicata Caput IV De Algebra ad Trigonometriam planam applicata	Volume V. Algebra applied to Elementary Geometry Section I: Algebra applied to Elementary Geometry Section II: Algebra applied to Trigonometry
Caput VI De Algebra ad Geometriam Sublimiorem applicata	Algebra applied to Sublime Geometry

Elementa Matheseos Universæ, **Elementa Analyseos Mathematicæ, Pars Secunda**	
Sectio Prima De Calculo Differentiali; Sectio Secunda De Calculo Integrali, seu Summatorio	Volume VIII. Analysis of the infinities
Sectio Tertia De Calculo Exponential Sectio Quarta De Calculo differentio-differentiali Sectio Quinta De Arithmetica Infinitorum	Volume IX. Exponential Calculus, Differentio-differential Calculus and Arithmetic of the infinities.[26]

Volumes VIII and IX are of particular relevance since they deal with infinitesimal calculus at a time when this new calculus was little known in Spain. Volume VIII is composed of eight sections and Volume IX is composed of four sections. In the following more detailed, comparative table, it can be seen how faithfully they follow Wolff's text:

---

[26] "Tomo I Arithmetica; Tomo II Geometria; Tomo III Logarithmica, Trigonometria Plana, Secciones de la Esphera, y Trigonometria Espherica; Tomo IV Tabla de los senos, y tangentes; Tomo VII. Algebra aplicada a la Arithmetica: Seccion I: Definiciones è Hypotesis necesarias para la inteligencia de las operaciones analiticas. Seccion II: Del modo de aplicar el Algebra a los Problemas determinados de la Arithmetica; Tomo V. Algebra aplicada a la Geometria Elemental: Seccion I: Algebra aplicada a la Geometria Elemental. Seccion II: Algebra aplicada a la Trigonometria; Tomo VI. Algebra aplicada a la geometria sublime; Tomo VIII. Analysis de los infinitos; Tomo IX. Calculo Exponencial, Diferencio-diferencial y Arithmetica de los infinitos".

Wolff	Wendlingen
Sectio Prima De Calculo Differentiali	**Volume VIII. Analysis of the infinities**
Caput Primum De natura Calculi differentialis	Section I. Differential Calculus Algorithms
Caput II De usu Calculi differentialis in tangentibus curvarum determinandis	Section II. Differential Calculus applied to find tangents, subtangents, subnormals and solve other problems that depend on them
Caput III De usu Calculi differentialis in Methodo de maximis et minimis	Section III. On the use of the Differential Calculus when searching for both maximum and minimum quantities
Sectio Secunda De Calculo Integrali, seu Summatorio	Caput I De natura Calculi integralis
Caput II De usu Calculi integralis in Quadraturis curvarum	Section IV [quadratures]
Caput III De usu Calculi integralis in rectificatione Curvarum	Section V. On the application of the Integral Calculus to the rectification of curves
Caput IV De usu Calculi integralis in cubandis Solidis et dimetiendis Superficiebus eorundem	Section VI. On the use of the Integral Calculus to find the solidity of solids and their surfaces
Caput V De usu Calculi integralis in Methodo tangentium inversa	Section VII. On the use of the Integral Calculus to find the curves that correspond to the given tangent
Caput VI De usu Calculi integralis in logarithmorum doctrina	Section VIII. On the use of Integral Calculus in Logarithmics
	Volume IX. Exponential Calculus, Differentio-differential Calculus and Arithmetic of Infinities

Sectio Tertia De Calculo Exponentiali	Section I. On Exponential Calculus
Sectio Quarta De Calculo differentio-differentiali	Section II. On Differentio-differential Calculus
Sectio Quinta De Arithmetica Infinitorum	
Caput I De natura Arithmeticæinfinitorum	Section III. On Arithmetic of the infinities
Caput II De usu Arithmeticæinfinitorum in Geometria	Section IV. Infinitesimal Arithmetic applied to Geometry[27]

Volume VIII starts in Section I with an introduction placing infinitesimals on the base of the Differential Calculus. According to Wendlingen's first definition, it is possible to find the foundation of the new calculus:

> Differential Calculus is the method for finding the infinitely small quantities, which added together give all of it.[28]

The approach that Wendlingen follows is influenced by Wolff's work, which in turn is part of the Leibnizian school of Differential Calculus:

> Calculus differentialis est Methodus quantitates differentiandi, hoc est, inveniendi quantiatem infinite parvam, quæinfinities sumta datam adæquat.

---

[27] "Tomo VIII. Analysis de los infinitos: Seccion I. De los Algoritmos del Calculo Diferencial; Sección II. Calculo diferencial aplicado al modo de buscar las tangentes, subtangentes, subnormales y resolver otros problemas que dependen de estas; Seccion III. Del uso del Calculo diferencial cuando se buscan así las cantidades maximas como las minimas; Seccion IV [cuadraturas]; Seccion V. De la aplicacion del Calculo integral a la rectificacion de las curvas; Seccion VI. Del uso del Calculo integral para hallar la solidez de los solidos y sus superficies; Seccion VII. Del uso del Calculo integral para hallar las curvas que corresponden a la tangente dada; Seccion VIII. Del uso del Calculo integral en la Logaritmica. Tomo IX. Calculo Exponencial, Diferencio-diferencial y Arithmetica de los infinitos: Seccion I. Del Calculo Exponencial; Seccion II. Del Calculo Diferencio-diferencial; Seccion III. De la Aritmetica de los Infinitos; Seccion IV. Aritmetica infinitesimal aplicada a la Geometria."

[28] "Cálculo diferencial es el método de hallar cantidades infinitamente pequeñas, las cuales sumadas dan el todo." ([Wendlingen, 1756–1761, "Tomo VIII', "seccion I', §1,Cortes 9/2812, f. 2)].

([Wolff, 1713-1715, "Elementa Analyseos Mathematicæ tam finitorum quam infinitorum", "Pars Secunda", §1].)

Let us show a couple of examples of the exercises in this volume compared with the corresponding exercises in Wolff's text. The first belongs to the section on maxima and minima:

Wolff	Wendlingen
Problema XIX	Problem V
86. Super recta AB tanquam hypotthenusa triangulum rectangulum maximum construere.	64. Given a line, make a rectangular triangle so that it is the largest of all the triangles that can be made in this way and the line given is its hypotenuse.

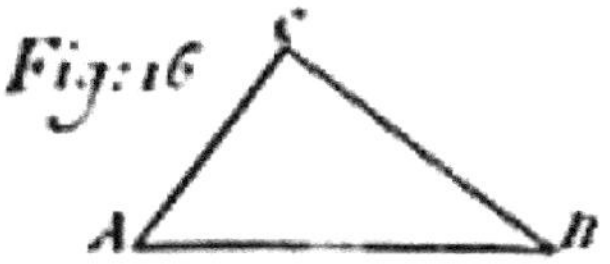

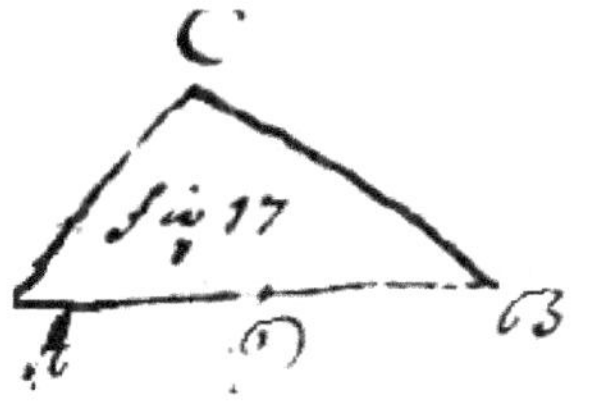

Figure 13. ([Wolff, 1713-1715, "Elementa Analyseos Mathematicæ tam finitorum quam infinitorum", "Pars Secunda", §86])

Figure 14. ([Wendlingen, 1756–1761, "Tomo VIII", "Seccion III", §64, RAH, Cortes 9/2812, f. 7])

Wolff:

Sit $AB = a, AC = x$, erit $BC = \sqrt{aa - xx}$, area $= \frac{1}{2}AC \cdot CB = \frac{1}{2}x\sqrt{aa - xx}$. Habemus adeo æquationem ad curvam tertii generis $x\sqrt{aa - xx} = 2y^2$, seu $aaxx - x^4 = 4y^4$.

Unde $2a^2xdx - 4x^3dx = 16y^3dy = 0, 2a^2x = 4x^3 \cdot \frac{1}{2}a^2 = x^2, \sqrt{\frac{1}{2}a^2} = x$.

Patet adeo triangulum maximum ess æquicrurum. Nam si $AB^2 = aa$ & $AC^2 = \frac{1}{2}aa$, erit etiam $CB^2 = \frac{1}{2}aa$; consequenter $AC = CB$.[29]

Wendlingen:

Resolution

Let the given line be $AB = a, AD = x$. It will be $DB = a - x$.

Let one of the Cathetus be $AC = x$. And it will be $\overline{AB}^2 = a^2, \overline{AC}^2 = x^2$.

Then $\overline{AB}^2 - \overline{AC}^2 = a^2 - x^2 = \overline{CB}^2$.

Then $CB = \sqrt{a^2 - x^2}$

---

[28]([Wolff, 1713-1715, "Elementa Analyseos Mathematicæ tam finitorum quam infinitorum","Pars Secunda", §86]).

And according to the nature of the problem it will be, $\frac{1}{2}x\sqrt{a^2 - x^2} = \sqrt{1/4a^2x^2 - 1/4x^4}$.

It then expresses a curve of the fourth kind.

Suppose it is equal to a maximum $= y^2$ it will be $\frac{1}{2}\sqrt{a^2x^2 - x^4} = y^2, \sqrt{a^2x^2 - x^4} = 2y^2, a^2x^2 - x^4 = 4y^4$.

$2a^2 xdx - 4x^3 dx = 16y^3 dy$.

That is $2a^2 xdx - 4x^3 dx = 0$, $2a^2 x = 4x^3, 2a^2 = 4x^2, \frac{1}{2}a^2 = x^2, \sqrt{\frac{1}{2}a^2} = x$.

That is, the largest triangle that can be made on the given hypotenuse is an isosceles triangle whose sides or cathetus must both be equal to the root of $\frac{1}{2}a^2$.[30]

We can see that although Wendlingen's exercise coincides with Wolff's text that of the Bohemian mathematician is a little more unclear. At the beginning he introduces a point D which he later does not use and does not explain that the expression he will later differentiate

---

[29]"Problema V. 64. Dada una recta formar sobre ella un triángulo rectángulo de modo que éste sea el más grande de cuantos se puedan formar de esta manera y que la línea dada sea su Hipotenusa. Resolución: Sea la línea dada $AB = a, AD = x$. Será $DB = a - x$. Sea uno de los Catetos $AC = x$. Y será $\overline{AB}^2 = a^2, \overline{AC}^2 = x^2$. Luego $\overline{AB}^2 - \overline{AC}^2 = a^2 - x^2 = \overline{CB}^2$. Luego $CB = \sqrt{a^2 - x^2}$. Y según la condición del problema será, $\frac{1}{2}x\sqrt{a^2 - x^2} = \sqrt{1/4a^2x^2 - \frac{1}{4}x^4}$. Luego expresa una curva del cuarto género. Supóngase ésta igual a un máximo $= y^2$ será $\frac{1}{2}\sqrt{a^2x^2 - x^4} = y^2, \sqrt{a^2x^2 - x^4} = 2y^2, a^2x^2 - x^4 = 4y^4, 2a^2 xdx - 4x^3 dx = 16y^3 dy$. Esto es $2a^2 xdx - 4x^3 dx = 0, 2a^2 x = 4x^3, 2a^2 = 4x^2, \frac{1}{2}a^2 = x^2, \sqrt{\frac{1}{2}a^2} = x$. Esto es, el triángulo más grande que puede formarse sobre la dada como hipotenusa es un triángulo equicruro [isósceles] cuyos lados o catetos han de ser iguales cada uno a la raíz de $\frac{1}{2}a^2$." ([Wendlingen, 1756–1761, "Tomo VIII", "Seccion III", §64, Cortes 9/2812, f. 7)].

corresponds to the area of the triangle.

The second example is from Section IV on quadratures:

Wolff	Wendlingen
120. *Hyperbolam Apollonianam intra asymptotos quadrare.* Quoniam ad Hyperbolam intra asymptotos $a^2 = by + xy$, seu si siat $a = b = 1$ (quod ponere licet, cum quantitatis b determinatio sit arbitraria, vi §. cit.) $1 = y + xy$, erit $1 : (1 + x) = y$, hoc est divisione actu facta, $y = 1 - x + x^2 - x^3 + x^4 - x^5 + x^6$&c. $ydx = dx - xdx + x^2dx - x^3dx + x^4dx - x^5dx + x^6dx$: &c.in infinit. $\int ydx = x - \frac{1}{2}x^2 + \frac{1}{3}x^3 - \frac{1}{4}x^4 + \frac{1}{5}x^5 - \frac{1}{6}x^6 + \frac{1}{7}x^7$ & *c.in infinit.*[31]	Problem XV 95. Make the quadrature of the Apollonian Hyperbola between the asymptotes. Resolution The equation which defines the Apollonian Hyperbola between the Asymptotes is $a^2 = bx + xy$. Suppose, as stated in the preceding Problem $a = b = 1$ and it will be $$1 = x + xy$$ $$\frac{1}{1+x} = y$$ This is actually to divide according to the method which I gave in algebra applied to arithmetic, when I taught how to divide quantities which do not themselves allow actual division. It will be $y = 1 - x + x^2 - x^3 + x^4 - x^5$& Then $ydx = dx - xdx + x^2dx - x^3dx + x^4dx - x^5dx$ & $\int ydx = x - 1/2x^2 + 1/3x^3 - 1/4x^4 + 1/5x^5 - 1/6x^6$ & [32]

---

[30] [Wolff, 1713-1715, "Elementa Analyseos Mathematicæ tam finitorum quam infinitorum", "Pars Secunda", §120)].

[31] "Problema XV. 95. Cuadrar la Hipérbola Apoloniana entre las asíntotas. Resolución.

La ecuación que define la Hipérbola Apoloniana entre las Asíntotas es $a^2 = bx + xy$.

Supóngase por lo dicho en el Problema antecedente $a = b = 1$ y será $1 = x + xy$, $\frac{1}{1+x} = y$.

Esto es dividiendo actualmente según el método que dí en el álgebra aplicada a la

Here we can see that Wendlingen's error is to calculate the "integral" from the same expression that appears in Wolff's exercise when the equation of the hyperbola considered by him has justly interchanged variables.

6   Criticism of the scientist or the Court official?

In 1760, Antonio Capdevila (1722–1778), a Catalan physician and mathematician, published a short text ([Capdevila, 1760]) in which he explained the mistakes found in Wendlingen's *Elementos*. Capdevila´s critical attitude towards Wendlingen is particularly sharp in a letter he to him a few months before the text was published and in which he reported that he had written this text enumerating Wendlingen's errors. This letter is part of the correspondence between Capdevila and Albrecht von Haller, a Swiss-German physician and scholar:

> [. . .] When in 1759 I read your *Matheseos elementa*, I was extremely happy [. . .]; but (according to what many of your compatriots wrote to me) it did not satisfy the requirements of the Germans. [. . .] I beg you to learn our very rich language [. . .], and read the *Elementa Matheseos* by your compatriot Christian Wolff very carefully, [. . .] as well the books by Spanish Mathematicians, [. . .].
>
> I am sending these corrections for you to take into account for the new edition of your *Elementa*.[33]

At the end of this letter Capdevila added: "Father Wendlingen's

---

aritmética, cuando enseñé como se dividen las cantidades que de por sí no admiten la división actual será $y = 1 - x + x^2 - x^3 + x^4 - x^5$&. Luego $ydx = dx - xdx + x^2dx - x^3dx + x^4dx - x^5dx$&, $\int ydx = x - \frac{1}{2}x^2 + \frac{1}{3}x^3 - \frac{1}{4}x^4 + \frac{1}{5}x^5 - \frac{1}{6}x^6$&" ([Wendlingen, 1756–1761, "Tomo VIII", "Seccion IV", §95, Cortes 9/2812, f.9]).

[33]"[. . .] Cuando el 1759 pude leer sus Matheseos elementa, me embargó tal alegría [. . .]; pero (segun me escribieron muchos compatriotas suyos) no satisfizo las expectativas de los germanos. [. . .] le ruego [. . .] que aprenda nuestro fecundísimo idioma, [. . .], y que lea con atención los Elementa Matheseos de su compatriota Christian Wolff, [. . .] además de los libros de Matemáticas españoles, [. . .] Le mando estas correcciones para que las tenga en cuenta en la nueva edición de sus Elementa". ([Barona et al., 1996, p. 73]).

ignorance and arrogance prevented him from mentioning them".[34]

Furthermore, in 1761, Lanz de Casafonda (1721–1785), a noble member of the Council of the Indies, severely criticized Wendlingen, defining him as an awful teacher. Specifically, he has to say about him:

> Father Wendlingen began to teach mathematics, speaking Castilian poorly, and attracted by this spectacle some very clever young men went to his classes, and although some even attended them for three years, none learned more than the principles of Arithmetic and Geometry, Father Wendlingen began to teach mathematics, speaking Castilian poorly, and attracted by this spectacle some very clever young men went to his classes, and althoug some even attended them for three years, none learned more than the principles of Arithmetic and Geometry, because the Father Professors have not gone beyond these principles in fourteen years, nor have they done any "Conclusión pública" or produced any textbooks on mathematics.[35]

This was written eleven years after Wendlingen's arrival at the Court of Madrid. Although it is said Wendlingen had not published anything, it is nevertheless known the *Elementos* were published between 1753 and 1756.

In any event, Capdevila and Lanz de Casafonda's criticisms reflect the feeling among the Spanish enlightened community that Wendlingen as a teacher fell far short of what required in Spain, and even thought they had a wide circulation, the *Elementos* contain many mistakes. Moreover, according to Valverde ([2007]), Wendlingen's unfocused activity failed to produce the expected results regarding astronomical goals. What is certain is that the estrangement between Wendlingen and the

---

[34]"El Padre Wendlingen calló a esto por su gran ignorancia y soberbia" ([Barona et al., 1996, p. 73]).

[35]"[...] empezó el dicho Padre a enseñar en un castellano chapurreado las matemáticas, y con la novedad concurrieron al aula mozos muy hábiles, y aunque algunos asistieron por espacio de tres años, ninguno aprendió más que los principios de la aritmética y geometría, porque no salieron de aquí, ni han salido en catorce años los Padres Catedráticos, ni han tenido ningunas Conclusiones públicas, ni aun sacado un curso siquiera de matemáticas" ([Laz de Casafonda, 1972, p. 65]).

circles close to power coincided with the arrival of new teachers: Rieger and Cerdà.

However, we have perhaps to consider these criticisms in the context of the wider criticism of Jesuit practice, which sometimes assumes xenophobic nature. In the eighteenth century, and not only in that century, it is only possible to understand the development of science when seen under the patronage of the political and economic powers of the country. The Society of Jesus was acting as a loyal provider of men of science to the different European Monarchies, and it is in accordance with this role that Wendlingen was sent to Madrid. It is therefore difficult to distinguish which role prevails in the case of Wendlingen; that of a man of science interested in astronomical discoveries and a mathematician spreading the new infinitesimal calculus, or that of a loyal servant of the Spanish Monarchy teaching the future king. The reason for the criticisms directed at Wendlingen is not entirely clear. Neither is it not clear whether they were equally provoked by the perceived inadequacy of a careless scientist, his astronomical observations and his teaching, or whether they were part of a campaign to disassociate him from the Monarchy for political reasons.

7 Concluding statements

Firstly, it is important to point out the significant influence that Wendlingen exerted at the Spanish court. Indeed, the seventeen years that Wendlingen spent at the Court of Fernando VI and Carlos III reflect the policy of the Spanish Monarchy to incorporate science as a matter of state. The position of Chief Cosmographer of the Indies, charged with the responsability for cartography and the mining exploitations in the American colonies also involved the added responsibility for creating the astronomical Observatory in Madrid, with the aim of forming part of the European astronomical network, which was regarded as an essential requirement for transoceanic navigation. If the task of Royal tutor is added to all these other responsibilities, it is evident how decisive Wendlingen's influence on the Crown might be; so decisive as to arouse enmity and opposition within the circles of the Spanish Monarchy.

Secondly, it is possible to affirm that Wendlingen contributed to the incorporation of the developments of Spanish astronomy to the network of European astronomical observatories, especially the observatory in Paris.

Thirdly, one may conclude that the publication of the *Elementos de la Mathematica* was an attempt to produce a common textbook of mathematics for the Spanish Court, despite the criticisms for the errors it contained. Of particular note is Wendlingen's interest in emphasising the practical approach of his text.

Fourthly, the unpublished treatises on infinitesimal calculus shows, on one hand, the growing interest in the new calculus among the Spanish enlightened society, and on the other hand, Wendlingen's effort to transmit the knowledge of this new calculus from Wolff's work.

Finally, whether due to strong anti-Jesuit pressure, or for reasons of xenophobia, or because of the Bohemian scientist's own mistakes, it is evident that by 1767 Wendlingen, both as a teacher and as a scientist had failed to meet the requirements expected of him on his arrival in Madrid in 1750.

Manuscripts

[Wendlingen, 1756–1761] Wendlingen, Johannes (1756-1761). *Elementos de Mathematicas*, Tomo VIII: Análisis de los infinitos; Tomo IX: Cálculo Exponencial, Diferencio-diferencial y Aritmética de los infinitos, Madrid: Royal Academy of History, Cortes 9/2812, 9/3811.

Printed Sources

[Backer et al., 1960] Augustin de Backer; Alois de Backer; C. Sommervogel; A. Carayon. *Bibliothèque des Écrivains de la Compagnie de Jésus, V. 8*. Éditions de la Bibliothèque S.J., Collège philosophique et théologique, Louvain, 1960.

[Barona et al., 1996] J. Lluís Barona; Xavier Gómez; Juan A. Micó; Amparo Soler. *La correspondencia entre A. Von Haller y Antonio Capdevila*. Universitat de València, Valencia, 1996.

[Berenguer, 2016] J. Berenguer. *La recepció del càlcul diferencial a l'Espanya del segle XVIII. Tomàs Cerdà: introductor de la teoria de fluxions.* PHD on History of Science guided by the doctor Mª Rosa Massa-Esteve, Universitat Autònoma de Barcelona, 2016 http://www.tdx.cat/handle/10803/367217, https://www.educacion.gob.es/teseo/mostrarRef.do?ref=1211406.

[Capdevila, 1760] A. Capdevila. *Correcciones de los elementos mathematicos del M.R.P. Juan Wendlingen de la Compañía de Jesús, [...] Para la fácil inteligencia de dichos elementos, para mis Díscipulos y aficionados a tan nobles Ciencias.* Antonio Serrano y Diego Rodríguez, Córdoba, 1760.

[De la Torre Ruiz, 2013] R. A. De la Torre Ruiz. Pedro Fresneda, cosmógrafo mayor del Real y Supremo Consejo de las Indias. *Estudios Jaliscienses*, 93, Colegio de Jalisco (México: 38-53, 2013. Lafuente, Antonio; Valverde Pérez, Nuria (2003) Los mundos de la ciencia en la Ilustración española.

[Laz de Casafonda, 1972] Manuel Lanz de Casafonda. *Diálogos de Chindulza.* Edition, notes and introduction by Francisco Aguilar Piñal. Cátedra Feijoo, Oviedo, 1972.

[López Piñero, 1969] J. M. López Piñero, José. *La introducción de la ciencia moderna en España.*, Ediciones Ariel, Barcelona, 1969.

[Navarro Brotóns, 2001] J. Navarro Brotóns. Scientific activity in Spain and the Role of the Jesuits (1680-1767). G. P. Brizzi and R. Greci (eds), *Gesuiti e Università in Europa (secoli XVI-XVIII). Atti del Convegno di studi (Parma, 2001).* ClUEB, Bologna, 2002.

[O'Neill et al., 2001] C. E. S. I. O'Neill; J. Mª S.I. Dominguez, *Diccionario histórico de la Compañía de Jesús.* 4 vol, Roma: Institutum Historicum, S.I. Universidad Pontificia de Comillas, Madrid, 2001.

[Pelzel, 1786] F. M. Pelzel, *Boehmische, Maehrische und Schlesische Gelehrte und Schriftsteller aus dem Orden der Jesuiten.* Im Verlag des Verfassers, Prague, 1786.

[Poggendorff, 1863] J. C. Poggendorff, *Biographisch-Literarisches Handwörter-buch (zur Geschichte) der Exacten Wissenschaften.* Verlag Von Johan Ambrosius Barth, Leipzig, 1863.

[Romano, 1999] A. Romano. *La Contre-réforme mathématique : constitution et diffusion d'une culture mathématique jésuite à la Renaissance (1540-1640),* École française de Rome, Rome, 1999.

[Sánchez Martínez, 2010] A. Sánchez Martínez. La institucionalización de la Cosmografía americana: la Casa de la Contratación de Sevilla, el Real y Supremo Consejo de Indias y la Academia de Matemáticas de Felipe II. *Revista de Indias*, LXX, 250: 715-748, 2010.

[Sellés *et al.*, 1988] M. Sellés; J. Peset; A. Luis y Lafuente. *Carlos III y la ciencia de la Ilustración.* Alianza Editorial, Madrid, 1988.

[Valverde Pérez, 2007] N. Valverde Pérez. *Actos de precisión: instrumentos científicos, opinión pública y economía moral en la Ilustración española.* Consejo Superior de Investigaciones Científicas, Madrid, 2007.

[Wendlingen, 1753–1756] J. Wendlingen. *Elementos de la mathematica.* (4 vol.). Joachin Ibarra, Madrid, 1753–1756 (http://bdh.bne.es/bnesearch/detalle/bdh0000134206).

[Wolff, 1713-1715] C. Wolff. *Elementa Matheseos Universæ.* Apud Henricum-Albertum Gosse & socios, Genevæ, 1743–1752 (https://archive.org/details/b30413163_0001/page/518/mode/2up).

[Wolff, 1728] C. Wolff. *Tabulæ sinuum atque tangentium tam naturalium quam artificialium.* Prostant in Oficina Rengeriana, Francofurt & Lipsiæ, 1728.

Teaching Mathematics at the Barcelona Royal Military Academy of Mathematics (1739–1803)

Maria Rosa Massa-Esteve

Departament de Matemàtiques. Universitat Politècnica de Catalunya, Spain

m.rcsa.massa@upc.edu

Abstract

Two interconnected and relevant changes in mathematics occurred between the middle of seventeenth century to the end of eighteenth century: the passage of analysis as a method to a discipline regularly taught in academies by the second half of the 18th century, and the transformation of "mixed" mathematics into "physic mathematics". We want to analyse the role of some contributions from Spain in these transformations of mathematics in eighteenth century.

After the War of Succession (1701-1714), the Spanish Bourbon Monarchy played a relevant role in scientific and technological developments by establishing several institutions to promote higher education for the officer corps, and mathematics constituted a pillar of Academy's curriculum.

In 1739, a Royal Ordinance established the contents of the course in mathematics to be taught in academies. This course, prepared by Pedro Lucuce (1692–1779), consisted of eight treatises with a total of approximately 2,200 pages long on the main fields of mathematics, including "pure" mathematics (arithmetic and geometry), and "mixed" mathematics (cosmography, statics,

This research is made with the support of the project PID2020-113702RB-l00: "Mathematics, Engineering, Heritage: new challenges and practices (XVI-XIX centuries)" of the Spanish Ministerio de Ciencia e Innovación. The author would like to thank Antoni Roca-Rosell, Mónica Blanco and Antonio Linero for their ideas and comments in this research on engineering education.

hydraulics, architecture, artillery, and fortification). In this chapter, we wish to analyse the treatment of mathematics in engineering education at Barcelona Royal Military Academy of Mathematics. The analysis of this contribution shows us that this transmission sometimes represents an appropriation of a known relevant text, but in other occasions introduces new methods and new perspectives in the development of eighteenth century mathematics knowledge.

1 Introduction

In the seventeenth and eighteenth centuries, many changes occurred in the practice of mathematics. An essential transformation in mathematics was the establishment of analytic methods and the use of symbolic language as a formal language in mathematics, so that the new language of symbols and techniques could be used to obtain new results in several parts of mathematics. In this period, new disciplines such as Algebra, Analytical Geometry and the Infinitesimal Analysis emerged, introducing new procedures. Thus, Practical Geometry, including Trigonometry and Logarithmic Calculus, underwent major developments throughout the eighteenth century ([Martínez-Verdú et al., 2023]). All these subjects proved their utility for solving problems in Natural Philosophy or Physics, in Architecture, and Military Fortifications. Military Engineering was a field *par excellence* for the application of this new mathematical knowledge.

In Spain, as in other countries, the origins of "scientific" engineering lay in the Army, where officers acquired their training informally. After the War of Succession (1701–1714), the Spanish Bourbon Monarchy played a relevant role in scientific and technological development by establishing several institutions to promote higher education for the officer corps, and mathematics constituted a pillar of the academies' curriculum.

Established in 1720, the Royal Military Academy of Mathematics of Barcelona represents a pioneer and original example of the organization of education for engineers. The aim of the Academy was to give a high level of training in mathematics to both young officers and experienced soldiers of the army. In 1739, a Royal Ordinance set up the rules of the

Royal Military Academy of Mathematics and established the contents of the course in pure and "mixed" mathematics to be taught in academies. The mathematical course prepared by Pedro Lucuce (1692-1779)[1] and taught for almost 40 years in this Academy, is our aim of research. In fact, the generation of engineers trained in this Academy enjoyed an excellent scientific and technical level, and its repercussion and influence can be appreciated in its civil and military works.[2]

There are several studies on the Academy that describe the operation of the institution, the type of education, and the books that were used; other researchers have focused on the Academy's engineers as builders and architects, or as military personnel. Recent studies have also analyzed the contents of courses taught at the Academy.[3] Recently, we published a text that valued the topics of the course taught by placing them within the European context of this type of education and characterizing the Academy as an institution of "transition" from an educational institution for private courses to a formal technical school ([Massa-Esteve and Roca-Rosell, forthcoming]). However, there is a lack of studies that analyze the rest of the content of the courses, overall mathematics, and place it within the historical evolution of mathematics at that time.

Therefore, our purpose is to reflect on the evolution of mathematics in this period as well as on the teaching of mathematics for engineers in eighteenth century Spain, by analyzing some mathematical contents of Lucuce's Mathematical Course taught at the Barcelona Royal Military Academy of Mathematics.

In the first section we present the historical evolution of mathematics in this period, presenting a survey of the changes in mathematical ideas

---

[1]Pedro Lucuce studied Canon Law at the University of Oviedo, but he left this career to join the Army in 1711, during the War of Spanish Succession (1705-1714). After the War he joined a regiment in Madrid where he had the opportunity to study Mathematics on his own initiative. In 1730, he was elected simultaneously member of the Corps of Military Engineers and of the Corps of Artillery. He chose the engineering corps and in 1736 joined the Academy of Mathematics of Barcelona.

[2]See [Muñoz Corbalán, 2004], [Vérin, 1993], [Galland Seguela, 2008], and [Massa-Esteve et al., 2011].

[3][Riera, 1975]; [Alcaide and Capel, 2005]; [De Mora and Massa-Esteve, 2009]); [Massa-Esteve et al., 2011]; [Massa-Esteve and Roca-Rosell, 2014]; [Massa-Esteve, 2014]; [Roca-Rosell and Massa-Esteve, 2020] and [Massa-Esteve and Roca-Rosell, forthcoming].

through mathematical courses that may have influenced Lucuce's Mathematical Course. In the second section, we focus on the creation of the Royal Military Academy of Mathematics and discuss a text containing the proposal for the contents of the Mathematical Course. Finally, we analyse and compare the contents of some mathematical courses with Lucuce's Mathematical Course.

2 Mathematical courses: pure and "mixed' mathematics

The mathematical courses constituted scientific and encyclopedic works that appeared during the seventeenth and eighteenth centuries and dealt with pure and mixed or "physic-mathematics". As mathematical courses were the main source of inspiration for Lucuce's course, we reflect on and present a brief survey of the evolution of the idea of mathematics since the Renaissance linked to engineering training.[4] Pure mathematics was divided into Arithmetic and Geometry, depending on whether discrete quantity or continuous quantity was studied. Mixed mathematics refers to those disciplines of the physical world that make use of pure mathematics. Initially, they were associated with Music and Astronomy (*quadrivium*), but soon their field expanded and included scientific and technical activities: Optics, Perspective, Cosmography, Architecture, and Engineering.[5]

Starting in the sixteenth century, mathematical disciplines began to include a new range of knowledge, probably as a result of the increase in scientific discoveries and technical advances, by means of which the principles of Arithmetic and Geometry could be progressively applied. Technical developments in military and artisan fields, as well as in scientific instruments, had increasingly made mathematical disciplines a universal tool.[6] For these reasons, sixteenth century works dealing with mathe-

---

[4]Many texts on algebra also emerged in the Renaissance. See [Rommevaux, et al., 2012].

[5]Christian Wolff (1679-1754) included a survey of the mathematical courses printed before his own course. Wolff states that one of the earliest published mathematical courses was Pierre Hérigone's course. [Wolff, 1713–1715, Vol. V, 3-21].

[6]L. Roberts, S. Schaffer and P. Dear published a number of papers to demonstrate that mathematics, science and technology have a variety of origins including the

matics also inevitably included the question of the number of mathematical disciplines and their classification, which today we refer to as modern mathematics.

Let us consider some examples of the above in works appearing in the sixteenth and the seventeenth centuries. Niccolo Fontana (1499/1500–1557), known as Tartaglia, in the preface to his translation of Euclid's *Elements* (1565), includes a page where he analyses different classifications of mathematics,[7] and gives his own classification: "pure mathematics" – Arithmetic and Geometry – and "mixed mathematics" – all the other parts.[8] Mixed mathematics incorporated those subjects in which quantity was recognized as mixed with matter, in contrast to pure mathematics, where quantity was dealt with separately from matter. In the seventeenth and eighteenth centuries the idea of introducing natural philosophy under the label of "physic-mathematics" became more popular.[9]

The main mathematical courses in the seventeenth century were those of Pierre Hérigone (1634/1637/1642), Milliet Dechales (1674-1690) and Jacques Ozanam (1693), whereas those in the eighteenth century were Tomàs Vicent Tosca (1707-1715), Christian Wolff (1713-1715) and Bernard Forest de Bélidor (1725).

As one of the earliest courses, we can cite that by Pierre Hérigone

---

artisan or craft tradition (see for instance [Roberts, Schaffer and Dear, 2007]).

[7]Tartaglia describes what he refers to as the "vulgar", that is to say the mathematics most people know, which includes Arithmetic, Geometry, Music, Astronomy, Astrology, Cosmography, Geography, Corography, Perspective, *Specularia*, the Science of Weights, Architecture and many others. He also recognises as usual the *Quadrivium* formed by Arithmetic, Geometry, Music, and Astronomy. He then goes on to point out that Luca Pacioli, in his classification, modifies the *Quadrivium* by adding Perspective, and states that therefore it would consist of five parts, or three parts if Music were removed. Tartaglia also cites Pierre d'Ailly's classification, which concludes that Music, Astronomy, and Perspective belong to mixed mathematics ([Tartaglia, 1565], Introduction).

[8]See [Brown, 1991], [Dear, 1995], and [Dear, 2011].

[9]Later in the eighteenth and nineteenth century, when natural philosophy and experimental physics (electricity, magnetism, heat, etc.) intervene, we speak of physic-mathematics in the sense not only of collecting experiences and observations, but of applying pure mathematics and mechanics, giving rise to mathematical physics and engineering sciences.

(1580-1643) published in Paris.[10] His *Cursus Mathematicus*[11] has the above-mentioned features insofar as it deals with different bodies of knowledge of pure and mixed mathematics. He considered pure mathematics divided, in agreement with the type of quantity, into Geometry, continuous quantity, and Arithmetic, discrete quantity; and mixed mathematics, where quantity is in conjunction with matter, divided into Optics, Mechanics, Astronomy, and Music. The *Cursus* consisted of five volumes and, later, six volumes. The first and second volumes dealt with pure mathematics (Geometry, Arithmetic and Algebra). The third and fourth volumes were devoted to mixed mathematics, that is, to mathematics used in fortifications or navigation. The fifth volume (the last one in the first edition, 1637) also dealt with mixed mathematics, including spherical trigonometry and music. In the second edition (1642), he added a sixth volume that dealt with Algebra, Perspective and Astronomy. As may easily be noticed, many different aspects of mathematics, as well as its applications, are represented in Hérigone's course.

We should mention next the three volumes of pure and mixed mathematics by Claude François Milliet Dechales (Chambéry, 1621-Turin, 1678),[12] which form the *Cursus seu mundus mathematicus* (1674).[13] This work included all the books (14) of Euclid, Arithmetic, the Spheres of Theodosius, Trigonometry, Practical Geometry, Mechanics, Statics, Universal Geography, Civil Architecture, and so on. This course was widely read and exerted much influence in Europe; for example, it influenced later Mathematical Courses like Tosca's *Compendio* and Lucuce's *Course*.

In eighteenth century Spain, following the tradition of the mathemat-

---

[10]On Hérigone's course see [Massa-Esteve, 2008] and [Mellado Romero, 2022]. Hérigone's date of birth is not known, probably 1580. He died in 1643.

[11]Hérigone, *Cursus Mathematicus*, 1634/1637/1642.

[12]Dechales was a Jesuit missionary in the region now called Turkey. For four years he gave public courses on Mathematics at the College of Clermont in Paris. Before teaching in Lyon and in Chambéry, he worked in Marseille, where he taught the arts of navigation and military engineering and practical applications of the mathematics to the sciences. From Marseille he moved to Turin where he was named professor of mathematics at the University. He died there at the age of fifty-seven. For more information, see [Schaaf, 1970-1991, pp. 621–622].

[13]It was republished in 1690, after the death of its author, with various new annexes to bring the contents up to date.

ical courses, Tomás Vicente Tosca (Valencia, 1651- 1723) wrote a *Compendio Mathematico* (1707-1715) consisting of twenty-eight treatises in nine volumes. Navarro-Brotons demonstrated that Tosca's *Compendio* is based on the work *Cursus seu mundus mathematicus* by Dechales.[14] Tosca's first volume deals with Elementary Geometry, lower–level Arithmetic and Practical Geometry; the second addresses higher Arithmetic, Algebra, and Music; the third, Trigonometry, Conic Sections and Machinery; the fourth, Statics Hydrostatics, Hydrotechnics, and Hydrometrics; the fifth, Civil Architecture, Military Architecture, Pyrotechnics, and Artillery; the sixth, Optics, Perspective, Catoptrics, Dioptrics, and Meteors; the seventh, Astronomy; the eighth, Practical Astronomy, Geography, and Seamanship; and the nineth, Gnomonics, the Ordering of Time, and Astrology.

Tosca's work went into three further editions during the eighteenth century, and some volumes were published separately. This is an indication of the reception and influence of this work. The first volume also contains a "Brief Introduction to Mathematical Disciplines" in which Tosca explains the object, nature and division of these sciences as they were understood at that time. Thus, he classifies them into "purely" mathematical sciences: Geometry, Arithmetic, Algebra, Trigonometry, and Logarithmic; and into physic-mathematical sciences: Music, Mechanics, Statics, Hydrostatics, Civil Architecture, Military Architecture, Artillery, Optics, Geography, Astronomy, and Cosmography. Tosca did not use the traditional classification of mathematics into pure and mixed, but rather classified it into pure and *physic-mathematics*. Throughout the seventeenth and eighteenth centuries, works employing this term in the title were published,[15] and the idea of introducing natural philosophy under the label of physic-mathematics became increasingly widespread.

Also in the eighteenth century, and as an antecedent to Lucuce's course, we may invoke the work of Bernard Forest de Bélidor[16] entitled

---

[14]For more information on Tosca's *Compendio,* see [Navarro Brotons, 1985].

[15][Dear, 1995, pp. 168-179].

[16]Bernard Forest de Bélidor was born in Catalonia in 1698, because his father belonged to the French Army fighting in the Nine Years' War (1688–97). He joined the Army and, given his knowledge of mathematics, he became a teacher at the new school of Artillery of La Fère (1720). In 1722, he became a member of the Académie des Sciences of Paris and in 1726, the Royal Society of London.

Nouveau Cours de Mathématique, à l'usage de l'artillerie et du génie ou l'on applique les Parties les plus utiles de cette science à la Théorie & à la Pratique des différents sujets qui peuvent avoir rapport à la Guerre (Paris, Chez Nyon, 1725; second édition, 1757). As the title indicates, it was an eminently practical course to be used as part of military training in which Bélidor was involved. This book was used in the new artillery schools established by royal decree. In fact, in the introduction, we can find a description of the composition and running of these artillery schools. The ten sections ("parties") of Bélidor's 1725 *Nouveau Cours* consist of:[17] 1) ten books including eight concerning Geometry and two concerning conics; 2) Trigonometry; 3) theory and practice of levelling; 4) measurement (*Toisé*) in general; 5) application of Geometry to the measurement of surfaces and solids; 6) application of Geometry to the Division of Fields; 7) application of Geometry to the use of proportional compass; 8) movement and fall of bodies (three books); 9) Mechanics (nine books); and 10) equilibrium and movement of fluids (three books). It is worth noting the explicit use of the expression "application of Geometry" in some of the titles and in the content lists of the sections. In the preface, Bélidor stated that one can consider the use of mathematical principles after using the theory in practice; in Bélidor's own words:

> Since it is only by applying Theory to Practice that one can make individuals aware of the applicability of numerous principles of which they do not see the utility otherwise, I have taken steps to make the mathematical principles relevant by attaching them to a number of different subjects of interest to engineers and artillery officers, as will be seen in what follows.[18]

Bélidor is considered the modern creator of engineering science.[19] His works were enormously influential during the eighteenth century in France and many other countries.

---

[17]The 1757 edition of the *Nouveau Cours* is completely renewed.

[18][Bélidor, 1725], Préface.

[19]Bélidor, *La Science des ingenieurs dans la conduite des travaux de fortification et d'architecture civile*, 1729. See [Vérin, 1993, Ch. IV].

In this sense, the promoters of the Royal Military Academy of Mathematics of Barcelona shared the general orientation of these authors and emphasized that engineering training should take place through a mathematical course based on these courses as we will show in the next section.

3 Mathematical Courses for the Royal Military Academy of Mathematics (1720-1803)

The Corps of Military Engineers (1711) as a technical organization was proposed by the engineer-in-chief of the Corps Jorge Próspero de Verboom (Brussels, 1667– Barcelona, 1744)[20] who had received mathematical training at a military Academy set up in Brussels by Sebastián Fernandez de Medrano (1646-1705).[21] Verboom followed the idea of his master and proposed the creation of military academies of mathematics, emphasized the usefulness of mathematics for the art of the war.

In 1720, a Royal Military Academy of Mathematics was established in Barcelona. The promoter was Verboom who proposed that the Crown created a network of academies with this objective. A document by Verboom, dated 1730, for the administration of academies (Verboom sought to create one academy in each province) provides information on his ideas of scientific contents in military education. According to Verboom, students needed theoretical and practical education for the profession of engineer in Geography, Mechanics, Optics, Architecture, Physics, Astronomy and more. Verboom claimed:

> Universal Geography, celestial and terrestrial, in particular, terrestrial and maritime; the three Architectures, Military,

---

[20] Jorge Próspero de Verboom was born in Brussels in 1667. He was educated in the Academy of Brussels directed by Sebastián Fernández Medrano. In 1692, he succeeded his father as an important engineer in the Low Countries. It seems that he met the French engineer Sebastien de Vauban. In 1709, he was in charge of the creation of the Spanish Corps of Military Engineers, approved in 1711. He died in 1744.

[21] This Academy of Brussels owed its existence to the personal initiative of this officer who was convinced that a military officer should have a complete military training. On the Academy of Brussels, see [Navarro Loidi, 2006].

Civil and Hydraulics; the moving forces and all kinds of Machinery compressive and elastic; Optics and Perspective; Conic sections from which all the above-mentioned architectonical subjects depended, the Science of Projections, planes and profiles, all of which made possible all types of buildings, constructions and machinery; the principles of Astronomy and Art of Navigation for the effective execution of all kinds of marine buildings; and finally the Theory and Experimental Physics of natural things, without such knowledge the universal and excellent profession of Engineer would not only be limited but also discredited ... [22]

In this document "Summary project or idea for the formation, government and permanent establishment of royal and military academies of mathematics and fortifications. . . ", Verboom also explained which kind of mathematics is necessary for engineers, in fact, mixed mathematics or mathematics with physical principles for the movement of the Stars and bodies, useful also for the winds. In Verboom's words:

The Dictation will consist of a treatise on Mixed Mathematics, namely, of all those that have Physical and Experimental Principles in which all the Dissertations previously given and concerning Natural Philosophy will be completed, namely, of the principles, Causes, Effects and Purposes of Natural Things, insofar as they are sensitive and experienced; Of the Elements and their Qualities; Of the Place and movement of the Orbs, Stars and Planets; From the vacuum of matter

---

[22]"La Geografía {Universal} celeste y terrestre, particular terrestre y Marítima; las tres Arquitecturas, la Militar, la Civil y la Hidráulica; Las fuerzas movibles y toda especie de Maquinaria Compresiva y Elástica; la Óptica y la Perspectiva; las Secciones Cónicas de quienes penden todos los aciertos {más ocultos y difíciles} de las citadas Arquitecturas, la ciencia de los Proyectos cuyos conciertos en planos y perfiles como por las deducciones escriturarias deben manifestar la posible ejecución de todo género de edificios, Construcciones y Máquinas; los principios de la Astronomía y de la Náutica para la competente erección de los Edificios Marítimos; y finalmente la Teórica y físicos experimentos de las cosas Naturales, sin cuyos conocimientos no sólo es limitada la universal y excelente profesión del Ingeniero, pero la desacreditan ..." ([Verboom, 1730, p. 18], "Proyecto o Idea", and the transcription in: [Roca-Rosell and Massa-Esteve, 2020, p. 176]).

and from the form of mobility and immobility of the opaque and transparent grave bodies; Of the sensitive, solid, hard, flexible and fluid qualities; Of the winds and meteors... Since this Science is the source from which those who aspire to the preeminent profession of Engineer have to avail themselves for the possible execution of their continuous and most important assignments.[23]

In this text from 1730, Verboom also indicated the sources that the director of the Academy had to use to write the course to be taught. For the first six books of Euclid, he suggested Port Royal's new "Elements" of Geometry and also the course prepared by Jean-Pierre de Crousaz (1663-1750), professor at the Lausanne Academy. In fact, in the seventeenth century, after the publication and dissemination of Descartes' *Géométrie* (1637), the relationship between algebra and geometry began to be considered from another perspective and the order of Euclid's *Elements* by teaching Geometry was questioned. It was also considered that there was an excess of propositions and that the Euclidean proofs were too long and difficult. The most representative texts of the Geometry of Port Royal mentioned by Verboom correspond to the *Nouveaux éléments de Géométrie* (1667- 2nd ed. 1683) by Antoine Arnauld (1612-1694), in the work *Eléments des Mathématiques ou Traité de la Grandeur in général qui comprend l'arithmétique, l'algèbre, l'analyse et les principes de toutes les sciences qui ont la grandeur pour object* (1680 – several more editions until 1765) by Bernard Lamy (1640-1715), and in *La Geometrie des lignes et des surfaces rectilignes and circulaires* (1718) by Crousaz. These texts established a new order of presentation for the

---

[23]"El dictado que compondrá de un tratado de matemáticas mixtas, a favor de todas aquellas que tienen principios físicos, y experimentales en que se completarán todas las disertaciones anteriormente dadas y concernientes la filosofía natural, a favor, de los principios, causas, efectos y fines de las cosas naturales en cuanto son sensibles y experimentadas. De los elementos y sus cualidades, del lugar y movimiento de los orbes astros y planetas, del vacuo de la materia y de la forma de la movilidad e inmovilidad de los cuerpos graves opacos y transparentes, de las cualidades sensibles, sólidas, duras, flexibles y fluidas de los vientos y meteoros, y de todos los movimientos de las aguas superficiales y subterráneas... Por ser esta ciencia el manantío del cual los que aspirasen a la preeminente Profesor de ingeniero, se han de valer para la posible efectuación de sus continuos y más importantes encargos... " (Verboom, AGS, Guerra Moderna, bundle 2994, 1730 and [Roca-Rosell and Massa-Esteve, 2020, p. 182]).

elements of Geometry, providing shorter demonstrations and new ideas. Regarding this new order of Port Royal, Verboom in his "Project" of 1730 explained:

> The first duty is to elaborate a regular Mathematics Course as compendious as possible which has, as its foundation, the first six books of Euclid in the natural, brief, methodical, easy, practical and demonstrative order that the gentlemen of Port Royal have manifested in their new elements of Geometry and that, later, Mr. Crousaz, professor at the Lausanne Academy, has carried out in his Geometry of the lines. All kinds of demonstrations of very few and brief general propositions, adapted to natural knowledge, are recognized to depend on both texts.[24]

In addition, regarding books 11 and 12 of Euclid's *Elements*, Verboom suggested the use of other mathematical texts such as Tosca's course. And on the other parts of mathematics, like the Practical Geometry, he proposed Bélidor's course. Verboom claimed:

> and regarding the 11th and 12th books of Euclid and other parts of Mathematics, observe the method of Father Tosca's course, avoiding the curious and abstract of his own demonstrations, which must conform to the method of the aforementioned Authors; and finally, as regards purely practical Geometry, and the practical Instructions of the Engineer and the Gunner, the said Royal Professor will use the Mathematical Course of Mr. Bélidor Royal Professor of one of the five Academies established in France for the same purpose...[25]

---

[24]"Primero debe elaborar un curso regular de mathemáticas el mas compendioso que sea posible que tenga por fundamento los seis libros de Euclides en el orden natural, breve, methódico, fácil, práctico y demonstrativo que ha manifestado posible los señores de Port Royal en sus nuevos elementos de geometría y después executado por el señor **Crouraz** [Crousaz] Profesor de la acadèmia de Lausana en su geometría de las líneas en cuyos ambos documentos se reconocen pender todos los géneros de demonstraziones de muy pocas y breves proposiciones generales acomodadas al conocimiento natural." (Verboom, Archivo General de Simancas, AGS, Guerra Moderna, bundle 2994, 1730, 24, and [Roca-Rosell and Massa-Esteve, 2020, p. 198].

[25]"y por lo tocante al 11º y 12º libro de Euclides y demás partes de Matemáticas

Therefore, the content of the Academy's teachings must be included in the tradition of these mathematical courses like Tosca's or Bélidor's.[26]

Despite the specific recommendations of Verboom in 1730, when later, in 1739, the Royal Ordinance that should guide the functioning of the Academy was published, no specific treatise was cited and it was only ordered that the director make the choice of the more useful mathematical treatises, writing the subjects as if they were their own. Thus, the *Ordinance* specifies:

9. To achieve education according to this idea, the Director General should select the most useful mathematical Treatises, arranging them methodically so as to be of greatest benefit to the Academics, writing the subjects that must be taught as his own doctrine, which should be (include) everything that is (to be) imparted in the Academy, extending explanation as far as should be deemed necessary.[27]

In the new Ordinance published in 1751, which ratified the earlier Ordinance of 1739 and modified it in some aspects, it was also established that the director would write the course as it were his own; in this case, the new Ordinance did not mention anything about using other courses.

---

observárase el método del Curso del padre Tosca evitando lo curioso y abstracto de sus propias demonstraciones las cuales deberán conformarse al método de los citados Autores; y finalmente, por lo que toca a la Geometría puramente práctica, e Instrucciones prácticas del Ingeniero y del Artillero, se servirá dicho Profesor Real del Curso Matemático del señor Belidor Profesor Real de una de las cinco Academias establecidas en Francia al mismo fin ... " . Verboom, Archivo general de Simancas (AGS), Guerra Moderna, bundle 2994, (1730), 44, and [Roca-Rosell and Massa-Esteve, 2020, p. 198].

[26] Also for the Architecture. Verboom quoted the use of geometry by Sébastien Leclerc (1637-1714). See [Roca-Rosell and Massa-Esteve, 2020, p. 198].

[27]"9. ...deberá el Director General elegir los Tratados más útiles de las matemáticas, ordenándolos con sucesivo método para el pronto aprovechamiento de los Académicos, escribiendo las materias que se han de dictar, como doctrina suya, que ha de ser cuanto en la Academia se explicare, extendiéndose en cada parte, según lo hallare conveniente." [Ordenanza, 1739], 8.

4 Treatises of mathematics in Lucuce's Course for Military Instruction

The first appointment for supervision of the Royal Military Academy of Mathematics in Barcelona was given to Calabro.[28] He was responsible for the preparation of a mathematical course, the content of which is described in his letter of 1724 to the Count of Montemar. In 1737, after a controversy between the director and Verboom, as a new director was named Pedro Lucuce.

Lucuce prepared an unpublished course that was taught for more than 40 years. Although the course was never published, it has been preserved in several manuscripts, written by different students from the Academy of Barcelona and from two other centers in Oran and Ceuta.[29] It contains a total of approximately 2,200 pages on the main fields of mathematics, including "pure" mathematics (Arithmetic and Geometry), and "mixed" or physic-mathematics (cosmography, statics, hydraulics, architecture, artillery, and fortification). In fact, at the beginning of the course, Lucuce explained the mathematics he intended to use and divided it into pure and not pure or mixed (he called them *impuras*). He also called the last group "physico-mathematics":

> Mathematics is divided into various parts, which are reduced to two types, one pure and the other mixed (impure); pure mathematics is Arithmetic and Geometry, because they deal with quantities regarding what is numerable or measurable; the others are physic-mathematics, because they consider quantities accompanied by some accident or perceptible attribute [*afección sensible*] typical of physics, such as optics, which deals with visible quantities and the music of the

---

[28] In fact, the directors of the Barcelona Royal Military Academy of Mathematics were: Mateo Calabro (1720-1738), Pedro de Lucuce y Ponce (1738-1756 and 1760-1779), Claudio Martel (1756-1760), Juan Caballero y Arigorri (1779-1784), Miguel Sánchez Taramas (1784-1789), Félix Arriete (1790-1793), and Domingo Belestá y Pared (1794-1802). See [Capel et al., 1998].

[29] We use a copy of Antonio Remón Zarco Torralbo y Orbaneja of 1759, located in the Central Militar Library, Madrid Ms. ML-R-235 (1759-1760). Zarco Torralbo was student of the Academy of Oran. See more details in [De Mora and Massa-Esteve, 2009] and [Massa-Esteve et al., 2011].

sound. Those to be studied in this Academy established for
the training of Military are.[30]

The name introduced by Lucuce to refer to mixed mathematics, like
Tosca in his *Compendio*, is physic-mathematics.

Therefore, the contents of Lucuce's *Mathematical Course for Military Instruction*[31] (1739-1744) consists of eight treatises. Treatise I:
on Arithmetic. Integers, linear algorithm; ratio and proportion usual
[*en común*]; rules of proportion; powers, roots, and progressions (270
pages). Treatise II: on elementary Geometry. Euclid's *Elements*; appendix on conic sections (331 pages). Treatise III: on Practical Geometry. Plane trigonometry, logarithms, construction of plane figures,
inscription and circumscription of straight figures in the circle, transformation of plane figures, use of instruments, planimetry, stereometry,
levelling (296 pages). Treatise IV: on fortifications. Regular fortification,
irregular fortification, effective fortifications in the field, campaign fortifications (315 pages). Treatise V: on artillery. The nature, composition,
inspection and conservation of gunpowder, ancient and modern artillery,
drawing of mortars, canons, gun carriages and weapons, canon and mortar batteries, fireworks, artillery trains and inventories of strongholds
(372 pages). Treatise VI: on cosmography. Celestial spheres, geography,
hydrography, seamanship, time measuring (392 pages). Treatise VII: on
statics; the motion of heavy bodies, machinery, hydraulics, optical compendium, general principles of optics, perspective (300 pages). Treatise
VIII: on civil architecture. Building decoration and beauty, the solidity
and safety of constructions (270 pages).

In the introduction to his Mathematical Course, Lucuce remarks
on the relevance of Aristotle's physics, which illuminates many other
natural sciences by means of mathematics, in accordance with the introduction of *Compendio*'s Tosca. He claimed as follows:

---

[30]"Dividiré la mathematica en varias partes, que se reducen a dos especies: la
una de puras y la otra de impuras: las puramente Mathematicas son la Arithmetica,
y Geometria, porque tratan de la cantidad en quanto numerable o mensurable; las
demás son Physico Mathematicas porque consideran la cantidad acompañada de algún
accidente o afección sensible propio de la Phisica, como la obtica que trata de la
cantidad visible, y la Música de la sonora. Los que se darán en esta Academia para
la instrucción de los Militares son." ([Lucuce, 1739–1744], s/n).

[31] *Curso Mathematico para la Instrucción de los Militares.*

The desire to know (said Aristotle) is natural to men, and among all the natural sciences that gives most satisfaction is Mathematics, for the lucidity of its truths, the energy of its proofs, and the clarity of its demonstrations. According to its Greek derivation, Mathematics is the same as Doctrine, or discipline, since it is exempt from the doubts and opinions so frequently found in the other sciences, for which reason whosoever devotes time to its study will not be disappointed, nor will the sweat shed in such an agreeable pursuit be in vain: therein the studious will find delight; since those fogs that often obscure the brilliance of other faculties do not cloud the sublime reaches of mathematics: rather such lights shine down from its soaring spheres to illuminate the pathways of the other natural sciences and lead to unmistakable truths.[32]

In the same introduction to his Mathematical Course, Lucuce also emphasized the usefulness of mathematics in physics, optics, astronomy, construction of buildings and organization of armies, among other subjects. In his own words:

With her (mathematics) the most closely hidden secrets of nature are revealed; she is the one that investigates the forces of impetus, the conditions of movement, the causes, effects and differences of sounds, the nature of light and its means of propagation. She raises buildings in beauty, makes cities well-nigh impregnable, orders armies with admiration and opens horizons to seafarers; lately she has ascended to the

---

[32] "Es natural en los hombres (dixo Aristoteles) el deseo de saber, y entre todas las ciencias naturales la que más le satisface es la Mathematica, por la limpieza de sus verdades, energía de sus pruebas, y claridad de sus demostraciones. Mathematica según la deribazion del Griego es lo mismo que Doctrina, o disciplina, por carecer de las dudas, y opiniones tan frequentes en las demás ciencias, por cuia razón no será malogrado el tiempo que se dedicare a su estudio, ni el sudor que se empleare en tan ameno campo: en el que experimentará el estudioso sus delicias; pues no llevan a la excelsa región de la mathematica aquellas nieblas que suelen oscurecer el resplandor de otras facultades: antes bien descienden de su remontada esfera tales luzes, que manifiestan las sendas a las otras Artes naturales para hallar la verdad con acierto". ([Lucuce, 1739–1744], s/n).

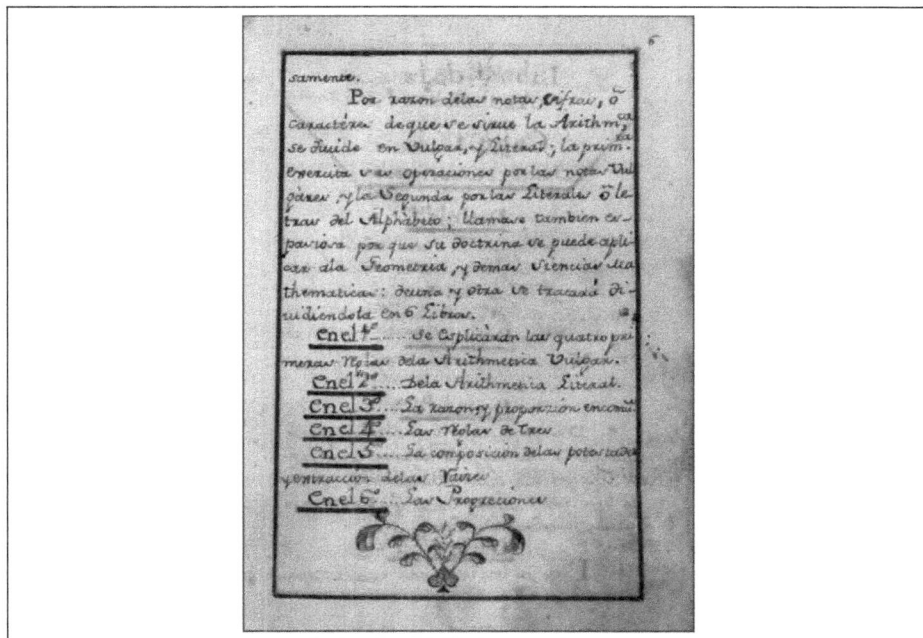

Figure 1: Introduction to the Treatise I ([Lucuce, 1739–1744, p. 6]).

heavens to chart the grandeur of the Stars and the harmony of their movements. . . [33]

After the introduction, Lucuce began Treatise I with the definition of Arithmetic as a science that dealt with discrete quantity, that is, numbers, and was divided into Speculative and Practical. Lucuce goes on to make a further classification and explains that, due to the notes, numbers or characters, Arithmetic is divided into vulgar and literal, the latter being treated by the literals or letters of the alphabet, also called "specious", because its doctrine can be apply to geometry and other

---

[33]"Con ella se descubren los más ocultos secretos de la naturaleza, ella es la que aberigua la fuerza del impetus, las condiciones del movimiento, las causas, efectos, y diferencias de sones, la naturaleza de la luz, y el modo de su propagación; levanta con hermosura los edificios, y hace casi inexpugnables las ciudades, ordena con admiración los exercitos, y abre camino a los navegantes; últimamente se remonta hasta el cielo para averiguar la grandeza de los Astros, y armonía de sus movimientos. . . "([Lucuce, 1739–1744], introducción, s/n).

mathematical sciences (see fig. 1). In his own words:

> By reason of the notes, numbers or characters that Arithmetic uses, it is divided into Vulgar and Literal; the first exercises its operations by the vulgar notes, and second by the literals or letters of the Alphabet; also called "specious" [referring to Viète's *speciosa*] because its doctrine can be applied to geometry and other mathematical sciences. Both [Arithmetics] will be addressed by dividing it into 6 books.[34]

As for Dechales and Tosca, the course begins with Geometry and then Arithmetic. On the other hand, Bélidor does not introduce Arithmetic in a separate treatise. So, Lucuce does not follow in his course the order of the mathematical courses quoted by Verboom.

Thus, Treatise I consists of six books: 1. The first four rules of the Vulgar Arithmetic (three chapters); 2. Literal Arithmetic (two chapters); 3. The ratio and proportion in common (in fact, Book V of Euclid's *Elements*); 4. The rule of three (four chapters); 5. Composition of powers and root extraction; 6. Arithmetic and geometric progressions (two chapters).

This first treatise of Lucuce's course has 280 pages. In it he explains both vulgar and literal Arithmetic and the Euclidean theory of proportions. Lucuce begins by affirming that Arithmetic is the door (according to Plato) to enter the knowledge of the other parts of mathematics, since they all need it. Arithmetic is the science that deals with discrete quantities or numbers and is divided into Speculative and Practical. The Speculative deals with the positions and properties of numbers and the Practical deals with their calculation or the four operations.

The entire Arithmetic treatise of Lucuce's course has a Euclidean presentation with definitions and propositions, which are sometimes theorems and sometimes problems, with corollaries and numbered scholia and numerical examples.

---

[34]"Por razón de las notas, cifras o caracteres de que se sirve la Arithmetica, se divide en Vulgar, y Literal; la primera exercita sus operaciones por las notas vulgares, y la segunda por las Literales o letras del Alphabeto; llamase también especiosa por que su doctrina se puede aplicar a la Geometría, y demás ciencias Mathematicas: de una y otra se tratará dividiéndola en 6 libros." [Lucuce, 1739–1744, p. 6].

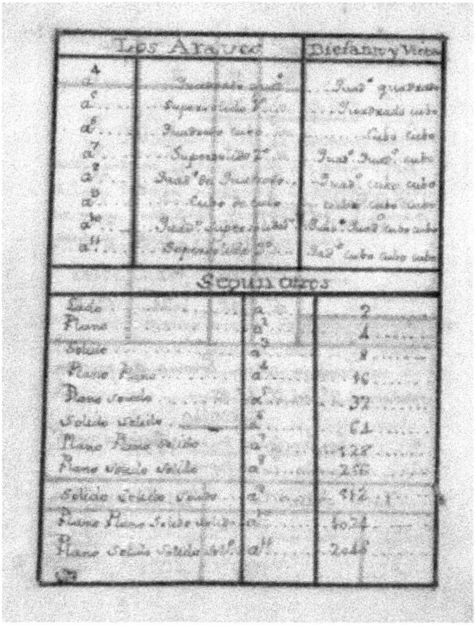

Figure 2: "Historical" Table of algebraic powers ([Lucuce, 1739–1744, Book 5, 102]).

In this treatise, we find some novelties with tables of powers with letters, and others quoting the notation from the Arabs, Diophantus and Viète (see fig. 2).

Moreover, we also find a table named "Synthetic Analytic", which reproduces the powers of binomials $(a + b)$ until the tenth power with letters.[35] Other authors made tables with numbers but not with letters. Lucuce in Book 6 made other table derived from Synthetic Analytic, which reproduces the geometrical means with letters (see fig. 3).

Now, we explain some ideas of Treatise II on Euclidian Geometry. In 1739, the reference in the Ordinance of the Royal Academy was that Euclid's *Elements* should form part of the syllabus ([Ordenanza, 1739], 8). However, Lucuce's choice to explain only Euclid's Books 1 to 6, and 11, 12 was usual at that time, because these books were considered to

---

[35]In Tosca (1707-1715) we find a similar table of powers of binomials with numbers, not with letters.

159

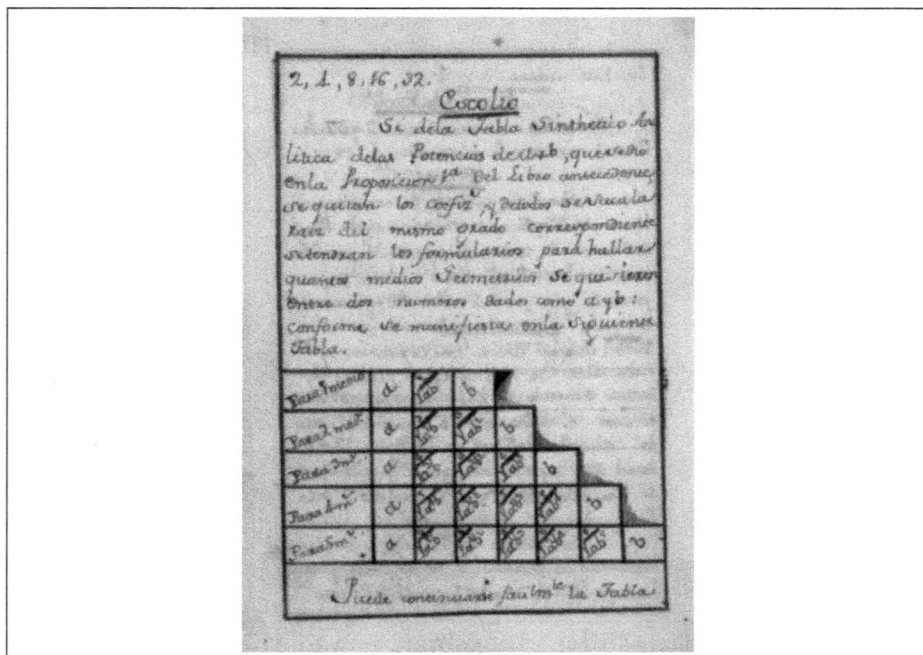

Figure 3: Table with letters of geometric means ([Lucuce, 1739–1744, Book 6, 138].

be the most useful.

In fact, Tosca in his *Compendio*, explains eight books, specifying that Books 7 and 8 correspond to books 11 and 12, respectively. Therefore, in his introduction to Treatise II, entitled "On Elementary Geometry", which deals with Euclid's *Elements* and Conic sections, Lucuce clarifies which books of the *Elements* he wishes to work with and in which treatise are explained. In an original and modern way, Lucuce shifts the book on the theory of proportions (Book 5) from the second treatise dealing with Euclidian Geometry to the first treatise dealing with Arithmetic, and Book 4 to the third treatise on Practical Geometry, perhaps following Tosca's course. In Dechales' and Ozanam's courses, all the books in Euclid's *Elements* remain together, while Tosca's course only shifts Book 4 to Practical Geometry. In Lucuce's words:

Since the work is extensive and diffuse, we explain in this

160

treatise (Treatise II) Books 1, 2, 3, 6, 11 and 12, with respect
to Book 4 (of Euclid's Elements) is addressed to Practical
Geometry (Treatise III) and Book 5 to Arithmetic (Treatise
I), while the others, being of little use, are omitted.[36]

Nevertheless, the originality of Lucuce resided in the fact that he preserved the Euclidian order, and less important propositions are omitted.
He emphasized this idea:

> The order I follow in the propositions is the same as that
> given by Euclid, so that they may be cited whenever neces-
> sary, the most useful being demonstrated with all possible
> brevity and clarity in order to save time for the explanation
> of other subjects that are of concern for the instruction of
> military personnel.[37]

Lucuce explained that in the books of the *Elements,* he provided
proofs using lines, letters, and numbers, thereby avoiding rhetorical discourse. Thus, in the introduction to Book 2 of the *Elements* in Treatise
II of the course, Lucuce stressed that the comprehension of these constructions with letters are useful for understanding Algebra (see fig. 4):

> In this book we consider the rectangles and squares which
> are formed by dividing a rectilinear line into parts; its com-
> prehension is of great usefulness in the Mathematics and es-
> pecially for Algebra; and, although its theorems are obscure,
> the demonstrations will be facilitated by lines, as well as by
> literal calculus and by numbers.[38]

---

[36]"Como la obra sea tan difusa y dilatada, explicaremos en este tratado los libros
1, 2, 3, 6, 11 y 12, respecto de que el 4° se dará en la Geometría Práctica, el 5°
se dio en la Arithmética; y los demás se omiten por ser de poca utilidad". [Lucuce,
1739–1744], Introduction.

[37]"El orden que seguiré en las proposiciones será el mismo de Euclides, para que se
puedan citar cuando convenga; demostrando las más útiles con la claridad y brevedad
posible; a fin de no malograr el tiempo, que se necesita, para la explicación de otras
materias propias también a la Instrucción de los Militares". [Lucuce, 1739–1744],
introduction.

[38]"En este libro se consideran los rectángulos y quadrados que se forman sobre una

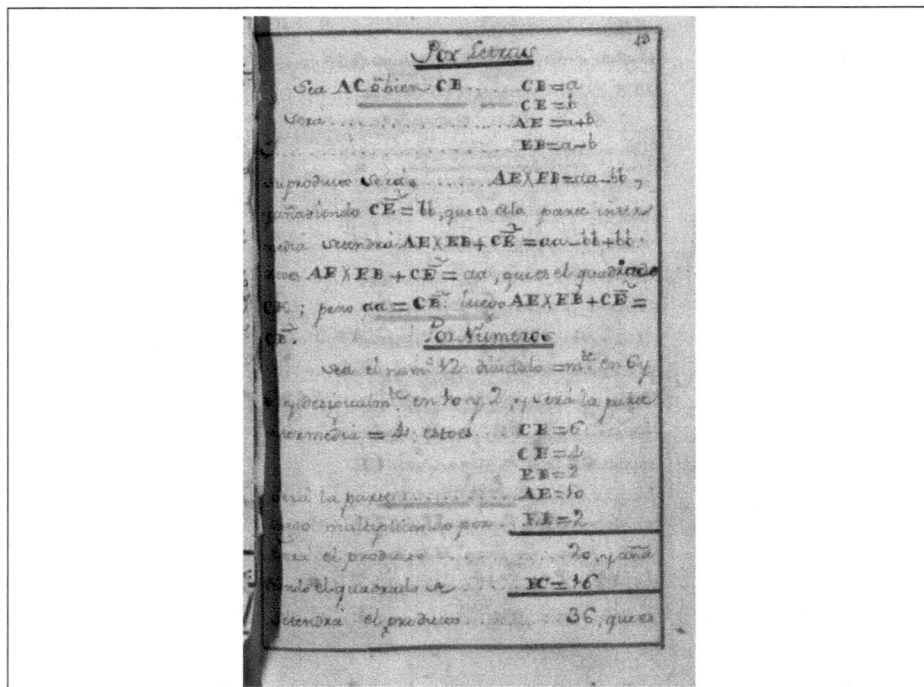

Figure 4: Lucuce's proof by letters and numbers. [Lucuce, 1739–1744], 49.

Tosca, in the same second book of the *Elements*, does the demonstrations with Geometry, only gives examples with numbers and does not present demonstrations with algebraic expressions.

As for Bélidor, in the preface where he describes the fourth book of the first part dedicated to Geometry, which deals with triangles and rectangles, he specifies that the first and second books of the *Elements* are summarized in 12 propositions and further claims that his fourth book contains more Geometry than the 62 propositions found in the *Elements*. As for the proofs, Bélidor uses algebra in some propositions (for example, in what is referred to today as the Pythagorean theorem),

línea recta dividida en partes; su inteligencia es de grande utilidad en la Mathemática y con especialidad para el álgebra; y aunque sus teoremas son obscuros se facilitarán sus demonstraciones assi por líneas como por el cálculo literal, explicándolas también por números." [Lucuce, 1739–1744], 42.

but he does so in a very different way from Lucuce.

Let us now consider the treatment of Book 5 on the theory of proportions, which is found in Treatise I of Lucuce's course on Arithmetic.

Lucuce performs an arithmetization of the Euclidian theory of proportions ([Lamandé, 2013]). Our aim is to characterize the arithmetization of theory of proportions that began in the XVI century according to Lamandé's article. In the first stage, some authors illustrate Euclid's definitions with numbers and/or algebraic symbols, whereas in the second stage, proportions apply to all kinds of quantities, both discrete and continuous. The third stage can be represented by the identification of a ratio with a numerical value. Finally, the proof of propositions on the theory of proportions is performed by arithmetic lemmas. Lucuce worked with all four stages of arithmetization: he illustrated with numbers and /or algebraic symbols, in addition he applied them to all kinds of quantities, he defined the ratio as the "exponent", and finally he used arithmetical lemmas in the proofs. Lucuce followed these stages of arithmetization approaching Euclid's *Elements* in an algebraic way.

Lucuce argued that the theory of proportions can be applied to numerical ratios as well as to discrete and continuous magnitudes, and that this is a necessary and universal key for comprehending mathematics as a whole, both pure and mixed. Thus, at the beginning of his Book III on the theory of proportions, in Treatise I, Lucuce explained the relevance and propaedeutic usefulness of the proportions as follows:

> Book 3 (Treatise I). On ratio and proportion in common. This book, which is the 5th by Euclid, deals with the ratio and proportion in common, whose doctrine is suitable for all kinds of quantities, both discrete and continuous, which serve for numbers, lines, surfaces and solids; being the universal key for acquiring the Knowledge of how many parts mathematics are composed.[39]

Once again, Lucuce specified that the propositions on the propor-

---

[39]"Libro 3°. De la razón y Proporción en común. En este libro, que es el 5° de Euclides se trata de la razón y proporción en común, cuia doctrina combiene a toda clase de cantidad ya sea discreta o ya contínua, esto es sirve para los números, líneas, superficies y sólidos; siendo llave universal para entrar en el conocimiento de quantas partes compone la Mathemática". [Lucuce, 1739–1744], 58.

tions explained in *Arithmetic* preserve the Euclidian order and are proven by letters and numbers, thus avoiding rhetorical explanations. He explicitly remarked that he presented only the propositions most useful. In Lucuce's words:

> His propositions preserve Euclid's, so they may be cited where appropriate, the (propositions) are less important and are omitted. Only the most helpful remain to be demonstrated by letters and explained by numbers in order to facilitate their understanding.[40]

Definition 5 on proportions presented by Lucuce is also original:

> Definition 5. The exponent of the ratio is the quotient obtained dividing the antecedent by the result. For example, if in the ratio of 6 to 2, 6 is divided by 2, the quotient 3 is the exponent that states the number of times that 6 contains at 2 and type $6/2$.[41]

Lucuce used this definition to show the relationship between ratios through the exponent of the ratio, that is, the numerical value of the ratio. In corollary 2, he states: "The value of the ratio will be exposed by the exponent."[42] Lucuce uses this definition to justify other definitions. In fact, Pedro Padilla (1724-1807?), a student of the Academy,[43] in his *Curso Militar de Matemáticas* (1753-1756) made identical developments.

---

[40]"Sus proposiciones guardan el orden de Euclides, para que puedan citarse cuando convenga; las menos principales se omiten y solo se darán las de mayor utilidad, que se demostrarán por letras, y se explicarán por números para facilitar su inteligencia". [Lucuce, 1739–1744], 58.

[41]"Definición 5. Exponente de la razón es el cociente que resulta dividiendo el antecedente por el consequente. Ejemplo si en la razón de 6 a 2 se divide 6 por 2, el cociente 3 es el exponente que declara el número de veces que el 6 contiene al 2 y se escribe $6/2$." ([Lucuce, 1739–1744], 59–60).

[42]"Corolario 2: El valor de una razón se expondrà por el exponente." ([Lucuce, 1739–1744], 60).

[43]The creation in 1732 of the Academy of Oran and in 1739 of that of Ceuta, following the ordinances of the Royal Academy of Mathematics of Barcelona, is relevant since Padilla was a student in Oran ([Capel et al., 1998, p. 126] and [Blanco and Massa-Esteve, 2018].

And Johannes Wendlingen (1715-1795), in his *Elementos de la Mathe-matica escrito para el uso de los principiantes* (1753), also presented the original definitions of exponent in his text referring to the ratio in a geometrical progression, possibly influenced directly or indirectly by Lucuce's approach.[44]

After the definitions and before the propositions, Lucuce presents two arithmetical lemmas that facilitate the demonstrations. Viète also used the first lemma to justify the solution to the construction of the second degree. Lucuce claimed:

> Lemma 1. If four quantities be proportional, the product of the two extremes is equal to the product of the two means; and if the product of the extremes is equal to the means, the quantities are proportional.[45]

Therefore, Lucuce used the idea of the exponent or the arithmetical lemma quoted above in the demonstrations of the propositions.

Lucuce's treatment of Euclidean Geometry, as we have shown, is different from that of the sources suggested by Verboom, both in the order of presentation of the books of the *Elements* and in the algebraic procedures for proving the propositions.

5   Some reflections

The scope of mathematics as a discipline is extended from the sixteenth century to the eighteenth century, when scientific matters became autonomous. The initial classification of mathematics in the sixteenth century, such as that by Pacioli or Tartaglia, were still derived from an "internal" analysis of the nature of mathematics and included a "mixed mathematics" in which quantity was linked to and mixed with other matters. At the beginning of the eighteenth century the classification of mathematics was extended with a new categorization changing the idea

---

[44]We find the same definition in [Lamy, 1692, p. 248] and in Wolff's work ([Wolff, 1713–1715, p. 44]).

[45]"Lemma 1º. Si cuatro cantidades son proporcionales el producto de los extremos es igual al de los medios; y si el producto de los extremos es igual al de los medios, las cantidades son proporcionales."[Lucuce, 1739–1744], 66.

of mixed mathematics and the name: physic-mathematics, justifying the introduction of subjects more related to natural philosophy or physics into the label of mathematics.

It is important to point out the emphasis placed on mathematics as the quintessential science for engineers at the Barcelona Royal Military Academy of mathematics. However, it should also be noted that, although the Academy instills a vision of mathematics that emphasizes the usefulness of the results, it respects the classical authorities, since it explicitly maintains the order and numbering of the definitions and propositions of Euclid's *Elements*.

Lucuce is presented to us as an original mathematician with his own ideas about teaching mathematics to train engineers. He constantly points out the didactic and utility criteria he uses. In addition, we have evidence of the high level achieved at the Academy thanks to a publication of 1755 dealing with the area of an oblique cone. The work shows a deep knowledge of analytic geometry and integral calculus by the engineers of Academy.[46]

Lucuce's course does not seem to have been prepared by making a direct copy of another course, but instead he selects content from other published mathematical courses and arranges them according to didactic criteria. In addition, Lucuce adds tables, interpretations, and proofs with algebraic language, omits the propositions that seem unhelpful, and changes the content to promote its understanding. In terms of pure mathematics, Lucuce's course does not follow the order of Bélidor's course or Tosca's course, which began with Geometry. In terms of contents, Tosca presents Arithmetic with many more propositions and parts that are not found in Lucuce's course. On the other hand, Bélidor does not include either vulgar Arithmetic, or the rules of three or company; he only includes, in the first part, a book of algebra before dealing with Geometry. As for Lucuce's Euclidian Geometry, he explains only some books of Euclid's *Elements* and treats Euclid's proofs by means of geometrical figures, numbers, and letters, thereby avoiding rhetorical discourse. We do not find them in Tosca's course, or in Bélidor's. The version of Euclid's *Elements* presented in Lucuce's course is explicitly adapted to his audience (military engineers and gunners).

---

[46]See the transcription of this work in [Cuesta Dutari, 1985, pp. 155-187].

Therefore, Lucuce presents an original course, designed with his own criteria for teaching engineers for almost 40 years at the Royal Military Academy of Mathematics in Barcelona, which takes into account new trends in science, but which is rooted in classical sources.

Manuscripts

[AGS Guerra, 2994] General Archive of Simancas (Archivo General de Simancas), Guerra, bundle 2994.

[Lucuce, 1739–1744] P. Lucuce. *Curso Matemático para la instrucción de los militares.* Madrid, Biblioteca Central Militar, First edition 1739-1744. There are the course notes of Antonio Remón Zarco del Valle, when he was cadet. Accessible in the web: https://bibliotecavirtual.defensa. gob.es/BVMDefensa/i18n/consulta/registro.cmd?id=4532 (accessed 30/11/2021).

[Verboom, 1730] J. P. Verboom. *Proyecto o Idea sumaria para la formación, gobierno y permanente establecimiento de Academias R[ea]les y Militares de Mathematicas y Fortificaciones ...*, Guerra Moderna, bundle 2994. Archivo General de Simancas (AGS), 35 pages.

Printed

[Alcaide and Capel, 2005] R. Alcaide and H. Capel. *El curso de cosmografía de Lucuce en las academias de Matemáticas Militares: el problema de los textos científicos y el desarrollo de la ciencia española del siglo XVII.* Geocrítica, Barcelona, 2005.

[Bélidor, 1725] B. F. Bélidor. *Nouveau Cours de Mathématique à l'usage de l'Artillerie et du Génie où l'on applique les parties les plus utiles de cette Science à la Théorie et à la Pratique des différents sujets qui peuvent avoir rapport à la Guerre.* Chez Nyon, Paris, 1725.

[Blanco and Massa-Esteve, 2018] M. Blanco and M. R. Massa-Esteve. La matemática pura en los cursos militares de matemáticas de Pedro Lucuce (1739-44) y de Pedro Padilla (1753-56). In D. Ruiz-Berdún, editor, *Ciencia y técnica en la Universidad.* Vol. II: pages 167–178. Servicio de Publicaciones de la Universidad de Alcalá, Alcalá de Henares, 2018.

[Brown, 1991] G. Brown. The Evolution of the Term 'Mixed Mathematics. *Journal of the History of Ideas*, 52/1 (Jan.-Mar.): 81-102, 1991.

[Capel et al., 1998] H. Capel, J. E. Sánchez and O. Moncada. *De Palas a Minerva. La formación científica y la estructura institucional de los ingenieros militares en el siglo XVIII*. Serbal, Barcelona, 1998.

[Cuesta Dutari, 1985] N. Cuesta Dutari. *Historia de la Invención del Análisis Infinitesimal y de su introducción en España*. Universidad de Salamanca, Salamanca, 1985.

[Dear, 1995] P. Dear. Art, Nature, Metaphor: the Growth of Physico-Mathematics. In *Discipline and Experience: The Mathematical Way in the Scientific Revolution*, pages 151-79. The University of Chicago Press, Chicago, 1995.

[Dear, 2011] P. Dear. Mixed Mathematics. In Harrison et al., editors, *Wrestling with Nature: From Omens to Science*, pages 149–172. University of Chicago Press, London, Chicago, 2011.

[Dechales, 1674] C. F. M. Dechales. *Cursus seu mundus mathematicus*. Ex Officina Anissoniana, Lyon, 1674.

[De Mora and Massa-Esteve, 2009] M. De Mora and M. R. Massa-Esteve. On Pedro de Lucuce's Mathematical Course: Sources and Influences. In H. Hunger, editor, *Proceedings of the 3rd International Conference of the European Society for the History of Science*, pages 835–844. Vienna, 2009.

[Galland Seguela, 2008] M. Galland Seguela. *Les Ingénieurs militaires espagnols de 1710 à 1803: étude prosopographique et sociale d'un corps d'élite*. Casa de Velázquez, Madrid, 2008.

[Lamandé, 2013] P. Lamandé. Quelques conceptions de la théorie des proportions dans des traités de la seconde moitié du dix-septième siècle. *Archive for History of Exact Sciences*, 67: 595–636, 2013.

[Lamy, 1692] B. Lamy. *Eléments des Mathématiques ou Traité de la Grandeur in général qui comprend l'arithmétique, l'algèbre, l'analyse et les principes de toutes les sciences qui ont la grandeur pour object*. Chez Henry Desbordes, Amsterdam, 3eme edition, 1692.

[Martínez-Verdú et al., 2023] D. Martínez-Verdú, M. R. Massa-Esteve, and A. Linero-Bas. Infinite Analytical Procedures for the Computation of Logarithms in Works by Benito Bails (1731–1797). *British Journal for the History of Mathematics*, online first: 1–34, 2023. https://doi.org/10.1080/26375451.2023.2186648.

[Massa-Esteve, 2008] M. R. Massa-Esteve. Symbolic language in early modern mathematics: the algebra of Pierre Hérigone (1580–1643). *Historia Mathematica*, 35: 285–301, 2008.

[Massa-Esteve et al., 2011] M. R. Massa-Esteve, A. Roca-Rosell, C. Puig-Pla. Mixed' Mathematics in engineering education in Spain: Pedro Lucuce's course at the Barcelona Royal Military Academy of Mathematics in the eighteenth century. *Engineering Studies*, 3 (3): 233–253, 2011.

[Massa-Esteve, 2014] M. R. Massa-Esteve. La Reial Acadèmia de Matemàtiques de Barcelona (1720-1803). Matemàtiques per a enginyers. *Quaderns d'Història de l'Enginyeria*, Vol. XIV: 17–34, 2014.

[Massa-Esteve and Roca-Rosell, 2014] M. R. Massa-Esteve and A. Roca-Rosell. Contents and Sources of Practice Geometry in Pedro Lucuce's Course at the Barcelona Royal Military Academy. In G. Katsiampura, editor, *Scientific cosmopolitanism and local cultures: Religions, ideologies, societies*, pages 329–335. National Hellenic Research Foundation/Institute of Historical Research/ Section of Neohellenic Research/ Programme of History, Philosophy and Didactics of Science and Technology, Athens, 2014.

[Massa-Esteve and Roca-Rosell, forthcoming] M. R. Massa-Esteve and A. Roca-Rosell. Teaching Engineers in Spain in the XVIII century. Analysing Mathematical Courses. In *ICHME Proceedings*. University of Mainz, Mainz, forthcoming.

[Mellado Romero, 2022] A. Mellado Romero. *La influencia del "Cursus Mathematicus" de Hérigone en la algebrización de la matemática*, PhD Thesis, Universidad de Murcia, Murcia, http://hdl.handle.net/10201/123566, 2022.

[Muñoz Corbalán, 2004] J. M. Muñoz Corbalán. *L'Acadèmia de Matemàtiques. El llegat dels Enginyers Militars*. Secretaría General Técnica del Ministerio de Defensa, Barcelona, 2004.

[Navarro Brotons, 1985] V. Navarro Brotons. *Tradició i canvi científic al País Valencià modern: 1660–1720: les ciències fisico-matemàtiques*. Eliseu Climent, València, 1985.

[Navarro Loidi, 2006] J. Navarro Loidi. *Las Ciencias Matemáticas y las Enseñanzas militares durante el reinado de Carlos II*. Ministerio de Defensa, Madrid, 2006.

[Ordenanza, 1739] *Ordenanza, e instrucción para la enseñanza de las Mathematicas en la Real, y Militar Academia, que se ha establecido en Barcelona*. Antonio Marín, Madrid, 1739.

[Riera, 1975] J. Riera. L'Acadèmia de Matemàtiques a la Barcelona Il.lustrada (1715–1800). In *Actes del II Congrès Internacional d'Història de Medicina Catalana*, pages 73–128, Barcelona, 1975.

[Roberts, Schaffer and Dear, 2007] L. Roberts, S. Schaffer and P. Dear. *The Mindful Hand: Inquiry and Invention from the Late Renaissance to Early Industrialization*. Koninklijke Nederlandse Akademie van Wetenschappen, Amsterdam, 2007.

[Roca-Rosell and Massa-Esteve, 2020] A. Roca- Rosell and M. R. Massa-Esteve. L'Acadèmia Militar de Matemàtiques de Barcelona. L'informe de 1730 de Jorge Próspero de Verboom. *Quaderns d'història de l'enginyeria*, XVIII: 159–208, 2020 https://upcommons.upc.edu/handle/2117/336655.

[Rommevaux, et al., 2012] S. Rommevaux, M, Spiesser and M. R. Massa-Esteve. *Pluralité de l'algèbre à la Renaissance*. Honoré Champions, Paris, 2012.

[Schaaf, 1970-1991] W. Schaaf. Dechales, Claude François Milliet. In C. C. Gillispie, editor, *Dictionary of Scientific Biography*, vol. 3: 621-2. Scribner's, New York, 1970-1991.

[Tartaglia, 1565] N. Tartaglia. *Euclide Megarense Philosopho, solo introduttore delle Scientie Mathematice. Diligentemente rassettato, et alla integrità ridotto, per il degno professore di tal Scientie Nicolo Tartalea Brisciano secondo le due tradottioni.* Curtio Troiano, Venetia, 1565.

[Tosca, 1757] T. V. Tosca. *Compendio Mathematico en que se contienen todas las materias mas principales de las Ciencias, que tratan de la Cantidad.* Imprenta de Joseph Garcia, Valencia, 1757, 9 v. First edition 1707-1715.

[Vérin, 1993] H. Vérin. *La Gloire des Ingénieurs: l'intelligence technique du XVI au XVIII siècle.* Michel Albin, Paris, 1993.

[Wolff, 1713–1715] C. Wolff. *Elementa Matheseos Universae.* Officina Libraria Rengeriana, Hale, Magdeburg, 1713-1715.

Algebra and the Teaching of Spanish Gunners during the 18th Century

Juan Navarro Loidi

Cátedra Sánchez Mazas UPV-EHU, Spain

jnavarrolo@gmail.com

Abstract

This chapter is about the changes that took place in the teaching of algebra to artillery officers in Spain during the 18th century. A huge change in the treatment of algebra is noticed in the books of artillery, and a less important but nevertheless considerable change in the treatises of mathematics for military education in that century. Three different periods may be distinguished in this process. For the first twenty years algebra was not included in artillery manuals, although it did appear in books on mathematics. During the second period, from 1720 to 1764, the Royal Military Academy of Mathematics of Barcelona was the leading institution in military theoretical training and algebra was also employed in artillery. Finally, from 1765 to the end of the century, the Royal Military College of Knights Cadets of Segovia monopolized the training of artillery officers, and not only algebra but also differential and integral calculus were taught at the College, leading to an improvement in the study of algebra. Furthermore, some resistance can be observed to the inclusion of algebra and especially of calculus in the curricula of the artillery academies, due to the lack of immediate applications to the gunnery of these mathematical structures.

This text is based on a paper presented at the symposium "Giants and dwarfs in the transformations of mathematics in the XVIII century" at the 26th International Congress of History of Science and Technology

1 Introduction

A study of the development of algebra in the 18th century reveals hardly any giant figure among the Spanish gunners or their professors. However, a gigantic change can be noticed in the role of algebra in the teaching of artillery officers during that century. This progression was not uniform and three periods can be considered:

- Until 1720, the institutions followed the achievements of the previous century. From 1701 to 1714, the kingdom was immersed in the War of Spanish Succession, during which the training of new officers was conducted in practice. Only some educational institutions, managed by Jesuits, continued to provide military theoretical teaching. When the war ended, several unsuccessful attempts were made to open new academies for cadets or officers.

- From 1720 to 1764, the training of the gunners was under the influence of the Royal Military Academy of Mathematics of Barcelona. This military institute was run by the engineering branch of the army, and the gunners were not happy about being taught by another arm. They opened some academies they believed to be controlled by themselves, but in practice they remained under the influence of the academy of the engineers. The Jesuits played only a small role in the theoretical instruction of army officers in this period.

- From the beginning of 1764, the teaching of gunners at the Royal Military College of Knights Cadets of Artillery of Segovia commenced with new guidelines and was freed from the influence of the Academy of Barcelona. The Jesuits were expelled from the kingdom in 1767 and the teaching for officers of king's artillery ceased to exist outside the College of Segovia [1]

---

[1] In this text, only the teaching of algebra to the artillery of the army is studied. Some schools or academies for naval gunners had existed in Spain since the 16th century and continued throughout the 18th century. These gunners followed a different tradition, and during this period the army and navy centres functioned separately. Although there was little difference in how algebra was taught in the training of officers in naval and army academies, the institutions, textbooks and professors were

The consideration of Algebra in each period will be studied by taking into account the curricula of the institutions, the mathematical treatises used in the courses and the books on gunnery published at that time.

2 From 1700 to 1720

King Charles II of Spain, of the House of Habsburg, died childless and subsequently a war broke out between the two pretenders to the throne, the French Philippe de Bourbon, supported by France, and the Austrian Karl von Habsburg, backed by Austria and other monarchies. It was eventually Philip who ascended the throne as Felipe V of Spain, but in the process the Spanish Empire lost its possessions in Italy and in the Low Countries.

Under the dynasty of the Habsburg House of Austria, military academies in the Spanish Empire up until 1700 were promoted either by the King, the Council of War, or under the direction of an army or other leading persons. The principal institution of this type in the late17th century was the Brussels Military Academy directed by Sebastián Fernández de Medrano.

Military arts were also studied in some Jesuit Colleges, in particular the Imperial College of Madrid, where fortification and artillery were sometimes given as independent subjects, or more frequently as a part of a course of mathematics. The best work that emerged from that teaching is the book *Escuela de Palas*.

In addition to those institutions, books were also written on gunnery by expert military men as well as sections devoted to artillery in some mathematical treatises; for instance, in the *Compendio Mathematico* (1707–1715) by the Oratorian Tosca. These books were used for self-learning or had an indirect influence on the training of the gunners.

---

completely different.The main institution for the navy was the Academy of Midshipmen of Cádiz, branches of which were later opened in Cartagena and Ferrol, and its principal professors of mathematics were Sánchez Reciente, Godin, Rovira and Ciscar, while Jorge Juan and Antonio Ulloa played a leading role in its history.

2.1 The Royal Military Academy of the Low Countries (1675–1704)

This academy was the leading centre in Spanish military teaching during the last quarter of the 17th century and the first decades of the 18th. It was opened in Brussels for officers belonging to the armies of the King of Spain. Its professor was Sebastián Fernández de Medrano, a self-educated second lieutenant who first fought on the border with Portugal and then from 1667 in Flanders. His academy had about twenty pupils a year. They were officers or cadets of the Spanish army, although many were of Italian or Flemish origin. Medrano's courses in Brussels included a first year devoted to fortification and practical geometry, and a second year for officers wishing to become military engineers who mainly studied Euclid's *Elements*. These advanced students helped Fernández Medrano with the classes for the new students.

Fernández de Medrano published many books on fortification, mathematics, geography and artillery for the academy or for self-learning. The principal books for each subject were *El Ingeniero* (1687), *Los seis primeros libros, onze, y doze, de los Elementos Geometricos del famoso Philosopho Euclides Megarense* (1688), *Breve descripcion del Mundo, y sus partes, ó guia geographica* (1686), y *El Perfecto Artificial, Bombardero y Artillero* (1699). He published no text on arithmetic or algebra.

2.1.1 Fernández de Medrano and Algebra

Medrano knew and appreciated algebra. In his version of Euclid's *Elements* he wrote about skipping books 7 to 10, as follows:

> I leave the seventh and the others, because it is a subject belonging to the incommensurable quantities that require the Algebra, which today can be found in a clearer and more understandable style, and more briefly, in what is called Specious or Analytic Algebra.[2]

---

[2] "El Septimo y los demas dexo, porque es cosa que pertenece a los grandores incommensurables que requieren la Algebra, la qual se halla hoy en estilo mas claro y comprensible y por mas breve camino, à que llaman la Algebra Speciosa ó Analitica" ([de Medrano, 1688, p. 266]. Translations by the author).

However, he neither taught algebra nor employed it in his books on mathematics, fortification, or geography, and even less so in those on artillery. It appears that he regarded algebra as being more important for mathematicians rather than for military men. Above all, when writing for gunners Medrano tried to avoid mathematics of any kind. In *El Practico Artillero* he stated that:

> Here I offer you (wise Reader) the *Practical Gunner*, stripped of confusion, as it is (for those who have not studied them) Arithmetic and Geometry[3].

Nevertheless, in his books on artillery he was obliged to explain what an angle is and how to measure it, or how to deal with weights and lengths, and therefore some mathematics was required to follow his books on artillery, but no knowledge of algebra.

In his books he used no algebraic symbols or formulas, not even when he added numeric examples to several propositions in Books II and V of his version of Euclid's *Elements*.

2.2 Escuela de Palas and algebra

Escuela de Palas (Milan, 1693) provides the best exposition of the military teaching imparted by Jesuits. The book contains eleven treatises on Arithmetic, Speculative and Practical Geometry, Plane Loci, Euclid's Data, Sphere, Geography, Numerous and Specious Algebra, Trigonometry and Logarithmic, the last and longest one on Military Arts. Although the author is anonymous, it was probably written by José Chafrión, a military engineer and disciple of the Jesuit José Zaragoza and of the Cistercian Juan Caramuel, with perhaps the help of the Viceroy of Milan, the Marquis of Leganés, another disciple of José Zaragoza. Some authors, for instance [Capel, 1988, pp. 101–102], have linked this treatise with a Spanish military academy that existed in Milan, which at the end of the 17th century was not operational.

---

[3] "Aqui te ofrezco (Lector prudente) el Practico Artillero, desnudo de confusion, como es (para los que no le han estudiado) la Aritmetica y Geometria" ([de Medrano, 1680, Prologue]).

The Preface of the book[4] is signed by Chafrion, who says:

> Since the Algebra both Numerous and Specious is the Miracle of Mathematics, the academic must apply himself to it with great attention, because there is no geometric problem, which is not solved by means of its Algebraic and Analytical rules, demonstrating whether its operation is possible, true, or fallacious.[5]

Algebra was therefore greatly appreciated by Chafrion.

The first ten mathematical treatises of *Escuela de Palas* are mostly shorter versions of courses given by the Jesuit Zaragoza. The book includes two different treatises on algebra, the reason for which is explained in the Prologue to "Numerous Algebra":

> The Analytical Art, or Algebra is divided into Arithmetic and Geometric; the first, since it executes its operations by numbers, is called Numerous Algebra or Major Art, while the second, since it encompasses both discrete and continuous quantities, and develops its resolutions by both numbers and lines, is called Specious Algebra.[6]

Both treatises share the same outline: firstly, definitions of expressions or polynomials and operations with them; secondly, polynomial fractions; thirdly, equations, and finally several problems solved by algebra. The author refers to Viète and Descartes in both algebras, and Cardano and Tartaglia only in numerous algebra.

---

[4]([Anonymous, 1693, *Exortación para los que quieren entrar é ser Discipulos en esta noble Escuela de Palas*]).

[5]"Siendo el Milagro de las Mathematicas el Algebra tanto la Numerosa, como la Especiosa, deve aplicarse a ella con mayor atencion el Academista, pues no ay Problema geometrico, que mediante sus reglas Algebricas, y Analiticas no se resuelva demostrando si es posible, verdadera, o falaz su operacion." ([Anonymous, 1693, *Exhortación*]).

[6]"Dividese el Arte Analítica, ó Algebra en Arithmetica y Geometrica, la primera porque executa sus operaciones por numeros se llama Algebra Numerosa, ó Arte Mayor; la segunda porque abarca tanto la cantidad discreta como la continua, y explaya sus resoluciones tanto por numeros, como por líneas se llama Algebra especiosa" ([Anonymous, 1693, first part, p. 167]).

The problems in Numerous Algebra are arithmetical and are solved numerically, while those of the Specious Algebra are of a geometric nature and are solved in three ways: with formulas, with numbers and with segments.

In the specious algebra, care is taken to provide homogeneous dimensions in the equations.

The symbolism and the names used in both treatises are different. For instance, in specious algebra the unknowns are called scalar quantities, and in numerous algebra cossic numbers. In specious algebra, the sixth power of the unknown is indicated by "ss" and in numerous algebra by "qc". The influence of Zaragoza is clear in Numerous Algebra, while Specious Algebra is closer to Viète's original texts than that in Zaragoza's book, but it is not known from where it has been taken. However, this book can hardly be original if the rest of the books on mathematics in *Escuela de Palas* are not originals.

The treatise deals mainly with mathematics and fortification, while artillery takes up only five pages in *Escuela de Palas*. There is no formula in the section devoted to artillery, neither are there any algebraic signs, not even + or −. The author follows Galileo on ballistics, but gives no formula to find the ranges, including only a table of the exact distances they reach.

2.3   Tosca and other authors

From the 16th century, the Spanish artillery opened practical schools to train gunners in working with cannons and gunpowder in strongholds or in operational armies. In these institutions, rough rules of thumb or specific instruments were proposed for quantitative or geometric questions. Sometimes concepts such as straight line, parallels, perpendiculars, polygons or circles were explained. One book for this elementary training published during the last decades of the 17th century is Fernández de Gamboa's *Memorias militares en que solo se trata del manijo de la Artillería*[7], or for the first years of 18th century *Exercicio militar*

---

[7]S. Fernández de Gamboa *Memorias militares, en que solo se trata del manijo de la Artillería*. Madrid, Mateo de Espinosa y Arteaga, 1671.

de artilleria[8], published in 1705 by S. López for naval gunners. There is no algebra at all in these books.

Furthermore, at that time, some treatises of mathematics included a book on artillery or pyrotechnics in the sections on applied mathematics. For instance, Milliet Dechales included one in *Cursus seu mundus Mathematicus*[9] and Christian Wolff in his *Elementa matheseos universae*[10]. In Spain, Tosca's *Compendio Mathematico*[11] contained a section on artillery. This treatise was very popular in Spain and ran to four editions in 1707–1715, 1727, 1757 and 1760, respectively. Although it was not intended for military education, it did have an indirect influence on it. Treatise V of the second volume is devoted to Algebra, and Tosca says at the beginning that:

> Algebra is therefore an Art that teaches how to find any quantity, solving the proposed question, by the same / terms, with which it was composed. From which is deduced that its object is more universal than that of Geometry, and Arithmetic, because contracting that to the continuous quantity, and narrowing this to the discrete, Algebra is extended without limitation to both, enriching them with new Theorems and Problems. [12].

---

[8]S. López *Exercicio militar de artillería compuesto por el capitan Simon Lopez, cabo de los artilleros de la Armada Real del Oceano y recopilado de muchos y graves autores*. Cádiz, 1705.

[9]C. F. M. Dechales *Cursus seu mundus Mathematicus*. Lyon, J. Anissonios, and C. Rigaud,1674. 3v. Pyrotechnics is in the 3rd volume.

[10]Ch. Wolff *Elementa matheseos universae*, 4 v., Halle, Libraria Rengeriana, 1713–1715. The section on Pyrotechnics is in the 2nd volume in this edition. First edition in German in 1710.

[11]Algebra is in the 2nd volume and Artillery and Pyrotechnics in the 5th ([Tosca, 1757]).

[12]"Es pues, la Algebra un Arte que enseña hallar cualquiera cantidad, resolviendo la cuestión propuesta, por los mismos términos, con que se compuso. De que se colige ser su objeto más universal que el de la Geometría, y Aritmética, pues contrayéndose aquella é la cantidad continua, y estrechéndose esta á la discreta, se extiende la Algebra, sin limitación á entrambas, á quienes enriquece con nuevos Teoremas, y Problemas." ([Tosca, 1757, T. 2, pp. 71–72]).

This treatise includes a chapter about operations with literal expressions, while others are devoted to the resolution of equations, with one or more unknowns and of the first degree or upper degrees. It also deals with irrationals and has a final chapter on applications of the algebra to the solution of linear or plane problems in geometry. The symbolism used in the book is almost modern. It differs from contemporary writing in the use of Ω instead of $=$ for equal, and in the way it indicates the exponent, which is written at the right of the base, but not as a superscript, a3 instead of a^3. In general, Tosca follows Milliet Dechales, but in this treatise he especially follows Rolle, Zaragoza, and the Spaniard Omerique for the applications of algebra to geometry.

The algebra is a quite up-to-date treatise, although in the section devoted to artillery there is no algebraic notation or formulas. Tosca believed that it was difficult to employ mathematics in artillery:

> Such is the condition of the fire . . . It is so sudden and rapid in
> its movement that it admits no mathematical examination,
> achieving almost total exemption of its precepts.[13]

However, he is of the opinion that mathematics can be used in ballistic to a certain extent, because for problems such as the precise ranges of a cannonball, the effect of air resistance and other reasons that may hinder the calculation of the trajectories can be avoided by taking as the maximum exact range the one obtained experimentally according to Blondel.[14] He consequently solves questions regarding angles of shot, or the horizontal range and maximum height of a shot, using arithmetic, geometry or trigonometry, but without formulas or algebraic notation.

2.4 The years immediately after the war of spanish succession

The Academy of Brussels was closed down in 1704, and the Spaniards and their French allies lost Brussels and Milan in 1706. The war prevented the army from opening new academies. Some teaching of military

---

[13]"Es tal la condicion del fuego [. . .] Es su movimiento tan pronto y veloz, que no tolerando examen mathematicc, casi logra total exempcion de sus preceptos." ([Tosca, 1757, T. 5, p. 427]).

[14]F. Blondel *L'art de jetter les bombes*, Paris, Langlois, 1683.

arts continued at the Imperial College of Madrid during the war, and the Jesuit José Cassani published *Escuela Militar de Fortificación* (1705). As its title indicates, the book is mainly about fortification, but it has an introduction containing elements of geometry, and also a section on artillery, but without calculations. Cassani generally follows *Escuela de Palas* in this book.

After the war, several requests were made to the King about the opening of academies for teaching military arts to officers and cadets. The Academy of Brussels was regarded as the example to follow in these matters, which were not very detailed and did not address programmes. The requests were confined to the study of the mathematics necessary for military arts, avoiding any speculative mathematical curiosities. It is important to point out that a branch of military engineering was created in the Spanish army in 1711. Engineers were more interested in mathematics than gunners at the beginning of this century, at which time the principal problem in the construction of a fortress was to protect every point of the walls with shots from at least another point of the fortifications, which can be reduced to a geometric problem. For its part, prior to Newton, gunnery had no a coherent theory to explain lived experience, and thus practice provided the rules for effective procedures, and the use of mathematics was limited.

Summarizing, before 1720 Algebra was generally known by the professors at the military academies or authors of treatises on artillery, although the teaching or use of algebra in gunnery was not regarded as necessary.

3 The royal military academy of mathematics of Barcelona

Around 1720, several attempts were made to open a military academy in different strongholds with the help of local military authorities, but many obstacles were encountered. In 1722, the King ordered the opening of academies for the training of artillery officers in Pamplona, Badajoz, Cádiz and Barcelona. These academies were dependent on artillery and had a short life span. The more long-lasting ones, Cádiz (1722–1727) and Barcelona (1725–1732), were confined to the teaching of some arithmetic and elementary geometry, but with little success. More successful was the academy opened by Mateo Calabro in Barcelona. He had tried to open an academy before, but it was not until 1720 that the project got under way. This academy was included in the Royal Order for the creation of academies for artillery officers of April 11th, 1722. At that time, Calabro was a lieutenant of artillery, but in 1723 he joined the corps of engineers, and his academy became dependent on that branch of the army. Consequently, the artillery opened a further academy in 1725, which turned out to be unsuccessful. In 1724, Calabro lodged a proposal with the commanders-in-chief of the army for a programme[15] for this academy. Although he did not include algebra as a subject, he proposed that "in the first class, literal arithmetic will be taught, and with it the main rules of numerical arithmetic will be demonstrated".

He taught symbolic writing, since in a preserved manuscript of his course on *Geometry* (1724)[16], the symbols $+, -, =$ are used, and for proportions, $A..B :: C..D$. In the absence of more data, one may assume that some algebra and symbolic writing was taught in his courses.

3.1 Pedro Lucuce and the *Reglamento ... de la real escuela o academia militar de mathematicas* (1739)

Pedro Lucuce replaced Calabro as the head professor of the Military Academy of Barcelona in 1738, and a *Royal Regulation* (1739)[17] was published that gave strong support to the academy. This military institute played a very important role in the training of officers in Spain during the 18th century.[18]

Until 1751, gunners were obliged to attend this academy in order to

---

[15]"Proyecto sobre el establecimiento formal de la Academia de Barcelona, dispuesto por el director de ella Don Matheo Calabro año 1724" ("Project about the formal establishment of the Academy of Barcelona, arranged by its director Don Matheo Calabro", year 1724).[AGS Guerra, 2994].

[16][Calabro, 1724]. Elementary and Practical Geometry by the succinct method taught in Barcelona in 1724 by Don Mateo Calabro ordinary engineer, and teacher of the school of artillery and navigation established in aforesaid capital.

[17][Ordenanza, 1739]. Ordinance and instruction for the teaching of Mathematics in the Royal and Military Academy, which has been established in Barcelona.

[18]A general study in [Capel, 1988]; for mathematics: [Massa-Esteve, Roca-Rosell, Puig-Pla, 2011].

receive theoretical instruction, because after 1732 no other academies of this type were available, and the institutions that did exist were very deficient. Lucuce wrote eight treatises for use in the classes taught at the Academy, and all the professors had to employ Lucuce's text when teaching this course. He did not provide the text in printed form, but the students had to keep clear notes of the subjects taught, and the course notes of some of the students are still preserved. These manuscripts give a good idea of what was taught at this academy. As far as Algebra is concerned, literal arithmetic, symbolic writing, formulas and some equations are explained in the treatise on Arithmetic.[19] The second chapter of this treatise deals with "literal arithmetic", and some symbolic writing is introduced; for instance $+, -, \times, =, >$. In the third book about proportions he introduces $\frac{a}{b}$ for fractions and $a..b :: c..d$ for proportions. He refers to Diophantus, Descartes and Viète, and for symbolism he prefers Descartes' notation to Viète's.[20]

In contrast to previous authors, Lucuce employs algebra in his Treatise on Artillery[21]. For instance, he deduces that F is in a parabola[22] from the fact that point F satisfies the formula $4YU \times NP = PF^2$. He proposes formulas to find the volume of the soil displaced by a mine, for instance $\frac{44a^3}{21}$ if the hole is cone-shaped.[23] For the number of cannonballs contained in a pile, he also obtains formulas[24] such as $\frac{mm+m}{2} \times \frac{2m+1}{3}$ or $\frac{mm+m}{2} \times \frac{m+2}{3}$.

It can be established from Lucuce's courses that basic algebra was taught and studied in Spain at the military academies in 1750.

3.2 The schools of mathematics for artillery in Barcelona and Cádiz (1751–1764)

New Schools of Mathematics for gunners were opened in 1751 in Cádiz and Barcelona. Their *Ordinance*[25] says that elementary and literal arith-

---

[19][Lucuce, 1759].

[20][Lucuce, 1759, f. 42 r.-v.].

[21][Lucuce, 1761].

[22][Lucuce, 1761, f. 124 v.].

[23][Lucuce, 1761, f. 148 r.]. He considers $\pi = \frac{22}{7}$.

[24][Lucuce, 1761, f. 185 v.-186 v.-r.].

[25][Ordenanza, 1751]. Ordinance and instruction to be observed in the Schools of Mathematics, which with the title of "Artillery" the King has ordered to be establish

metic had to be taught, among many other subjects, but without specifying any further details. The head of each school, or First Professor, had to set out the programme of his school, but in the absence of a general common supervisor, differences existed between the programmes in Barcelona and Cádiz.

The programme proposed by Juan Ramón Silby in 1752 for the School of Mathematics for the Artillery of Barcelona[26] is known. In the first treatise it includes the *Elements* of Euclid and some of Archimedes' theorems, but with the recommendation "to use from the 5th book on, whenever appropriate, the specious, symbolic, or algebraic method". This first treatise includes also at the end a section on the operations of the "vulgar and literal Arithmetic". The following treatises included arithmetic, practical geometry, civil architecture and fortification, artillery, and mechanics. In the last treatise, in addition to optics, perspective, and navigation, the "upper arithmetic or Algebra up to and including square equations" are incorporated.

No manuscripts of the courses given at this academy have been found. Nevertheless, a brochure, "Certamen Académico Mathematico"[27], published in 1754 to make a contest held with the best students of the academy known to the public, reveals a poor knowledge of the contemporary mathematics taught at the faculty. Among the most recent authors mentioned are Caramuel, Tosca and Fernández de Medrano; that is, Spaniards from the end of the 17th century, and for the best treatises *Escuela de Palas* and Tosca's *Compendio Mathematico*. It is said that Tosca's *Compendio* was the only complete course of Mathematics printed in Europe, which in 1754 is nonsense. It doesn't mention any foreign mathematician, even not Wolff, Leibniz or Newton, and it praises Spanish mathematics, which was far from being outstanding at that moment.

Gabriel Martínez's programme propsed in 1756 for the School of Mathematics for the Artillery of Cádiz is also known[28] It contained

---

in the Strongholds of Barcelona and Cádiz, under the direction of that branch of the Army.

[26]Treatises to be given and coordination of the same in the school of Mathematics established by order of H.M. of October 21, 1751 ([Silby, 1754]).

[27][Certamen, 1754].

[28]Report on the current state of the Royal School of Artillery of Cádiz and the

conic sections "demonstrated by algebra", powers and roots with "the operations that are made when imaginary roots occur", and a "sufficient instruction of the integral and differential calculus for the applications that are henceforth made in the other treatises". Quite reasonably, Mechanics was given "with the help of analysis and integral and differential calculus".

No manuscripts of the courses given in Cadiz have been found. In a public contest[29] in Cádiz, a question was set about "the differential calculus (problems of maximums and minimums), resolution of algebraic equations (even third degree) and formulas for areas and volumes of solids." Therefore, these courses taught in Cádiz by Martínez not only included algebra, but also integral and differential calculus, which signified an important advance.

The drawback of the curricula at these academies was their dependence on the capacities of the head professor, a position that changed hands quite frequently. In Cádiz, Martínez died, and in Barcelona the course became weaker, so at the end of the decade the programme followed by artillery was similar to that of the *Academia de Mateméticas de Barcelona*.

3.3 Military instruction by the jesuits (1720–1764)

Nothing about military arts was published by Jesuits in Spain between 1705 (Cassani) and 1764 (Cerdá). However, some artillery was taught at a small number of colleges, as may be deduced from the "Conclusiones", printed with the questions posed in public contests and organized by the Jesuits with the best students of these colleges.

In a contest in 1734, directed by Fr. Alvarez[30], and in another in 1744 by Fr Terreros y Pando[31], there is a section on artillery, but only

---

course that the ordinary Commissioner D. Gabriel Martínez has composed for its establishment with the maxims that have been practiced under his direction and subsequently carried out ([Martínez, 1756]).

[29] Description of the fifth and sixth thorough examinations, held at the Royal School of Mathematics under the care of the General Artillery Corps, in the stronghold of Cádiz, on May 24 and 25, 1754 ([Escrutinio, 1754]).

[30] [1734] about artillery the section entitled "De la Pirotechnia o Arte Tormentaria", [Álvarez, 1734, pp. 29–50].

[31] [Terreros, 1744]. Artillery in the section "Tormentaria", pp. 56–67.

practical questions were set for the students on this subject.

In 1751, again under the direction of Terreros y Pando[32], the questionnaire included algebra, "sublime geometry" (conics with equations) and differential calculus, while the questions on artillery concern characteristics of cannons and mortars without calculations. In 1760, with Bramieri[33] presiding, the questions on artillery refer only to parabolic trajectories, and in 1762 with Benavente[34], the situation was the same. Jesuits taught algebra at their colleges as part of mathematics, but it was not taught for artillery.

4 The new *Regulation* (1762) for artillery and the college of cadets of Segovia

As a result of a proposal from Count Gazola[35], King Carlos III published a new *Regulation* ([Reglamento, 1762]) for the Spanish artillery where it is stipulated as follows:

> XX In each Department a practical school of artillery will be founded for the instruction of officers and troops, . . . and since the Department of Segovia should have the best and most orderly instruction, not only for the gentlemen cadets, but also for all officers and troops, in addition to the practical school another theoretical school shall be established there for which the precise and suitable persons shall be named as professors.[36]

---

[32][Terreros, 1751]. Artillery is in the section " § IX ArquitecturaMilitar", questions 4 to 7, p. 17.

[33][Bramieri, 1760]. Gunnery in sections "Theorica del arte de arrojar bombas" and "Práctica del arte de arrojar bombas", [Bramieri, 1760, pp. 107–110].

[34][Benavente, 1762]. Ballistics in section "De las leyes del movimiento de los cuerpos aplicadas al tiro de las bombas", [Benavente, 1762, pp. 16–17].

[35]Felix Gazola or Felice Gazzola (Piacenza 1698–Madrid 1780), count Gazola, was a noble Italian who was head of the artillery of Carlos III when he was King of Naples and again when he became king of Spain. A multitalented man, he was an experienced general who fought in many battles against Austria and Portugal and also an archaeologist who run the digging of Paestum (Naples) or a decorator who designed the Throne Room in the Royal Palace of Madrid.

[36]"XX En cada Departamento se formará para instrucción de los oficiales, y tropa

The College of Segovia opened in 1764, and consequently the Schools of Céldiz and Barcelona were closed down.

4.1   Tomás Cerdá's *Lección de artillería* (1764)

In regard to the establishment of this new military college, the Jesuit Cerdá published his *Lección de Artillería* (1764), dedicated to Count Gazola, the promoter of the College of Segovia. This book constitutes a step forward in the scientific teaching of artillery. It consistently follows Newton's physics and uses fluents and fluxions ([Berenguer, 2020]).

For trajectories, he limits himself to the parabolic trajectories in order to avoid "many tangled calculi" needed if air resistance is taken into account. For internal ballistics, he poses some differential equations and for the initial velocity of a bullet he arrives at the formula:[37]

$$v = \sqrt{\frac{4}{3}\frac{rna}{x} - \frac{2}{3}n}.$$

On mining, he calculates the amount of soil displaced by a mine by integrals[38]

The book includes an "Appendix: Rules for counting cannonballs or bombs in artillery stores"[39], in which he gives the usual formulas and refers the reader to his book on algebra for the demonstrations, where the problem is studied quite deeply[40]. He cites Daniel Bernouilli and John Muller, but mainly follows Newton and the Englishmen B. Robins and F. Hawksbee, who applied Newton's theories to artillery. For practical questions, he mentions the French authors Belidor and La

---

una Escuela Práctica de Artillería ... y como el Departamento de Segovia ha de tener por objeto la mejor y más arreglada instrucción, no solo de los Cavalleros Cadetes, sino también de todos los Oficiales, y Tropa destinada en él, se establecerá allí además de la Escuela Práctica otra Theorica, para la que se nombrarán por Profesores los sujetos precisos y propios para su desempeño". [Reglamento, 1762, pp. 9–10].

[37][Cerda, 1764, pp. 74–103]: "Capitulo Cuarto. De la velocidad de la bala al salir del cañon, en que se trata de la fuerza de la pólvora." The formula can be found on page 99.

[38][Cerda, 1764, pp. 119–160]: "Capitulo quinto de las cargas y efectos de las minas".

[39]"Apéndice Reglas para contar las balas o bombas en los almacenes de artillería" ([Cerda, 1764, pp. 160–172].

[40][Cerda, 1758, pp. 210–223].

Valière, although the practical aspects of canons and mortars and how to fire them, and how to set up a battery, are not well developed and the book was not used at the College.

4.2 The Royal Military College of knights cadets of artillery of Segovia

The College[41] was opened in 1764, and the subjects to be studied there are outlined in its *Ordinances* (1768), as follows:

> The main ones [are] Calculus, Geometry, Mechanics, Hydraulics, Hydrostatics. Fortification and Artillery with some elements of other faculties[42]

These imprecise indications had to be specified by the First Professor. At the beginning this post was held by the Jesuit Antonio Eximeno, who had the help of lieutenant Vicente de los Ríos for the artillery.

During the early years, Eximeno and de los Ríos tried to introduce an updated programme, but were unable to establish it. Eximeno, like every other Spanish Jesuit, was expelled from the kingdom on April 2nd, 1767, and De los Ríos was principally working for the Royal Spanish Academy before he died in 1779.

After Eximeno's departure only basic algebra was taught at the College from 1767 to 1772, with equations up to the second degree, as in Lucuce's course. The course on mathematics improved when Cipriano Vimercati (1772–1777) was appointed First Professor. It contained eight volumes: the 1st and 2nd on arithmetic; the 3d and 4th on geometry; the 5th on algebra; the 6th on the application of algebra to geometry; the 7th on infinitesimal calculus and the 8th on mechanics. A manuscript with the notes taken by the cadet Felipe Silva on the course of algebra taught in 1774 has been preserved. It has nine chapters dealing with operations with numbers and letters, powers and roots, equations until 4th degree and indeterminate equations. Newton, Leibniz, Mac Laurin, Boscovich are mentioned, but he mainly follows Reyneau and Wolff.

---

[41] A general study in [Herrero, 1990]. For mathematics, see [Navarro-Loidi, 2013].

[42] "Los principales [son] el Cálculo, Geometría, Mecánica, Hidráulica, Hidrostática, Fortificación y Artillería con algunos elementos de las demás facultades" ([Ordenanza, 1768, p. 51]).

4.2.1 Giannini's Course

With his successor, the Italian Pedro Giannini, First Professor from 1777 to 1803, a regular course of mathematics was established at the College. He published *Curso Matemático* (4 volumes) and *Practicas de Geometría y Trigonometría con las tablas de Logaritmos* (1784) to be used as handbooks in the courses.

Curso.[43]   Volume I has three parts on elementary geometry, trigonometry and conic sections. The geometry is a version of Euclid's *Elements*.

Curso.[44] Volume II is on algebra and it will be commented on later.

Curso.[45] Volume III is divided into four parts: the first on the basis of differential calculus and the calculation of differentials and integrals; the second on the integrals of rational and irrational expressions; the third, differential equations of the first order, and the last one on differential equations of a higher order.

Curso.[46] Volume IV has three parts on statics, hydrostatics and dynamics. Giannini begins with Bernoulli's "Principle of virtual work", which gives a very mathematical version of mechanics.

[Giannini, 1782, Curso Matemático Tomo II] is about algebra, even if Giannini prefers to call it "universal arithmetic". He says in the prologue that:

> Universal Arithmetic is the object of the second Volume of this Course. It is divided into three Books: the first deals with the Algorithm of the numerical and literal quantities; in the second the Resolution and Construction of Equations is given, and finally in the third the use of universal Arithmetic is addressed through the Resolution of different Arithmetic and Geometric Problems.[47]

---

[43] [Giannini, 1779].

[44] [Giannini, 1782].

[45] [Giannini, 1795].

[46] [Giannini, 1803].

[47] "La Aritmética universal es el obgeto del segundo Tomo del presente Curso. Esta va dividida en tres Libros: en el primero se trata del Algoritmo de las cantidades numéricas y literales: en el segundo se dà la Resolución y Construcción de las Equaciones: finalmente en el tercero se hace ver el uso de la Aritmética univer-

In the prologue, Giannini recommends for the first book the study of Barrow and principally of Maseres and Castiglione.[48] The second book on equations contains the way to solve equations up to the 4th degree, and the study of curves corresponding to expressions of those degrees, including conics, cissoids, conchoids and many other curves, the resolution of equations in general and also series and their sums.

In this prologue, he has this to say about the contents of that second book:

> These theories treated in the best Algebra Courses composed by Messrs. Newton, Pietro de Martino, Clairaut, Maclaurin, Saundersons, P. Reynaud, Wolff, Mrs. Gaetana Agnesi, Vincenzo Riccati and Geronimo Saladini, Simpson, Bezout, Bossut, Leonhard Euler, Abbé Sauri, Caravelli, and others, have been reduced to general principles and demonstrated more easily.[49]

The problems in the third book are quite elementary at the beginning and are solved by the rule of three or first-degree equations. Giannini tries to apply them somewhat to artillery, including some questions regarding troops on the march, shots from several cannons or the distribution of stored material. Later on, there are more difficult problems about sequences and the sums of their elements, as well as some questions related to artillery; for instance, how to determine the number of cannonballs in a pile.

Very few references to applications of the theory to gunnery are found in the 4th volume. Equations and differentials and integrals are used to

---

sal por medio de la Resolución de diferentes Problemas Aritméticos y Geométricos." ([Giannini, 1782, Prologue]).

[48]He did not explain these recommendations. Barrow is mentioned by Giannini probably to emphasise a geometrical foundation of algebra. Maseres must be Francis Maseres (London, 1731– Reigate, 1821) a popular English author of books on algebra and Castiglione Johann Castillon or Castiglione (1704 Castiglione (Tuscany) – 1791 Berlin) publisher of an enlarged edition of Newton's *Arithmetica universalis*.

[49]"Estas teorías tratadas en los mejores Cursos de Algebra compuestos por los Señores Newton, Pedro de Martino, Clairaut, Maclaurin, Saundersons, P. Reineau, Wolf, Doña Cayetana Agnesi, Vicente Riccati y Gerónimo Saladini, Simpson, Bezout, Bossut, Leonardo Euler, Abate Sauri, Caravelli, y otros, han sido reducidas á principios generales y demostradas con mayor facilidad." [Giannini, 1782, Prologue].

solve the problems of a body moving under constant forces or under centripetal forces, or of the movement with resistance of the medium, but in an abstract way that is far removed from practice.

In general, Giannini's course consists of abstract mathematics rather than a course on mathematics applied to artillery. However, it is necessary to point out that the most difficult parts, such as that on differential equations in Curso III or on movements with resistance of the medium in Curso IV, were not explained in class, as may be deduced by comparing the content of these books with the course notes[50] taken by the cadet Tomás Eslava in 1781.

4.2.2 Morla's *Tratado de artilleria* (1784–1786)

Tomás Morla was the successor to Vicente de los Ríos as professor of artillery. He wrote his *Tratado de artilleria* 3v. with a fourth volume containing plates (1803).

While it is a book that deals predominatly with practical artillery, it also addresses theoretical questions using algebra, but avoiding calculations. For instance, when speaking about the force of exploding gunpowder, Morla states that: "Although on the subject of this article many very solid and exact theories have been written after the famous Robins"[51], he does not develop the subject in this treatise because it is too hard to deal with in a non-specialized book.

Regarding the trajectories of projectiles, he has a similar opinion and believes that practice is the only way to achieve accurate cannon shots, but not: "blind and servile practice divested of principles and theory"[52]

On mines, he gives five different hypotheses about the shape of the hole produced and discusses whether he should rely on calculations made with these hypotheses or on experimental data. Some calculations are made in this section by using roots and logarithms, and Morla proposes the use of a proportional compass by those who are unfamiliar with these parts of mathematics.

---

[50]T. Eslava *Elementos de los cálculos diferencial è integral.* 1781. Manuscript in Biblioteca Pública Yanguas y Miranda, Tudela de Navarra.

[51]"Aunque sobre el objeto de este artículo se haya escrito después del célebre Robins muchas teorías muy sólidas y exactas" ([Morla, 1784, p. XXVI]).

[52]"Práctica ciega y servil destituida de principios y teoría" ([Morla, 1785, p. 343]).

Sometimes he gives formulas, such as for the calculation of piles of cannonballs, where he explains that they are based on the theory of progressions, but provides no demonstration or any explanation where one can be found.

Generally he explains the subjects, justifying the hard points with references to Robins, Antony, Arcy or Newton, and sometimes to Bezout or Euler for mathematical questions. He makes no mention of Giannini, even though the cadets have studied with his books; this is because Morla considered his treatises too abstract. In 1796, in a report[53] to count Colomera, Director of Spanish Artillery, Morla and Autran, both generals of artillery, criticise Giannini and his courses saying that: "he is a mathematician and is neither a military man nor a gunner", and, among other criticisms, "his course has two major flaws: one gives much abstract and sublime theory of no or very little connection with these two professions, and the other lacks the continuous examples and applications to the military branches that should be in such a course". They conclude by stating that "the course of mathematics should not be for astronomers but for gunners."

4.2.3 The discussion about the curriculum

Not only Giannini's course was found to be too abstract, but also the syllabus of mathematics at the College of Artillery, which was considered too demanding by many members of the staff at the Institution. In December 1781 a School Council was devoted to discussing the programme. All its members except Giannini said that it would be better: "if the course was less abstract and compact; if it broadened its practical applications to become more accessible and easier to understand, and at

---

[53] [AGS Guerra, 5759]. This document is in a Report of count Colomera to the head of the army in answer to a request for an increase in salary for Giannini: "Informe del conde de Colomera" dated 10 February 1796. The quotes in Spanish: "que es matemático y no es militar ni artillero ... Que su curso tiene dos defectos capitales: uno dar muchas teorías abstractas y sublimes de ninguna o poquísima conexión con dichas dos profesiones y otro carecer de las continuas aplicaciones y ejemplos que debiera haber a los ramos militares en un tal curso" and "el curso de matemáticas no debe ser para astrónomos sino para artilleros".

the same time omitted sublime theories and advanced calculations."[54]

They asked for the existing syllabus to be replaced by one less difficult, without calculus or algebraic geometry, adding that "it remains to be seen whether all these theories expressed and calculations are essential for the training of an artillery officer."[55]

The head of the artillery refused these changes and backed Giannini. He wanted advanced mathematics to be included in the curriculum, even if the applications were not evident, but did not justify his decision. However, it was in line with the dominant ideology of the Enlightenment, which asked for noble and wise officers, and with the fact that the artillery was taking the control of the production of arms and gunpowder production in the kingdom, and well qualified artillery officers were needed to fill the managerial posts at these factories and to improve the army's weaponry.

5 Conclusion

Algebra was incorporated into the training of Spanish gunners during the 18th century in a quite long process. Fernandez de Medrano avoided the use of algebra in his books on artillery at the beginning of the century, while Lucuce used algebraic notation and some literal arithmetic in his Course on artillery, and Morla at the end of the century supposed algebra known by the majority of his readers. This evolution matched with the changes that were made in the mathematics taught to artillery officers. At the beginning of the century taught mathematics was mostly geometry, even for a learned author such as Chafrion; at the end algebra had an important place in the Course of Giannini in Segovia.

This important transformation was done not by a giant, but for many dwarfs, professors of the academies, gunners or Jesuits, who adjusted the progresses made by Leibniz, Wolff, Newton and others to the teaching of artillery in Spain.

---

[54]Minutes of the meeting of 3 January 1782): "si el curso fuese menos abstracto y conciso; si se extendiere en aplicaciones prácticas que lo hicieran más ameno y comprensible, y al mismo tiempo se omitieran las teorías sublimes, y cálculos superiores" ([Actas, 1765-, f. 373 r.]).

[55]"resta saber si son esenciales para formar un buen oficial de artillería las expresadas teorías y cálculos" ([Actas, 1765-, f. 373 r.]).

The extent of that evolution can be attributed to several causes. Military training in Spain underwent progress throughout this century, and the theoretical basis of gunnery was improved thanks to the application of Newton's theories to ballistics and to the development of Chemistry. The enlightenment of the Spanish governments, especially under Carlos III, eventually supported a more theoretical approach.

However, this change faced some resistance due to the absence of immediate applications to the Art of War, his lack of confidence continued until the late 18th century, mainly in regard to differential calculus, the application of which was difficult and whose usefulness was not entirely clear. As Morla said in his treatise on the equations of motion:

> This method of finding the precise range of bombs . . . is absolutely useless in the practice of the battery and consequently must be omitted in the schools. No matter how well-versed he may be in higher calculations, what officer in a battery would the time and proportion to form integral series and approximate them by repeating this operation in each barrel of gunpowder and to each mortar?[56]

To summarise, algebra was incorporated into the instruction of artillery officers at the end of the 18th century. Elementary algebra was well accepted, and algebraic geometry and integral and differential calculus were also admitted, despite the questions raised about the lack of evident applications, largely thanks to the support of the enlightened heads of the army.

Manuscripts

[AGS Guerra, 2994] General Archive of Simancas (Archivo General de Simancas), Guerra file 2994.

[AGS Guerra, 571] General Archive of Simancas, Guerra file 571.

---

[56]"Este medio . . . es absolutamente inútil en la práctica de la batería y de consiguiente debe omitirse en las Escuelas ¿Qué oficial podría por más instruido que está en los cálculos superiores tener en una batería, tiempo y proporción para formar series integrales y aproximarlas / repitiendo esta operación a cada barril de pólvora y a cada mortero?" ([Morla, 1785, pp. 342–343]).

[AGS Guerra sup., 11] General Archive of Simancas, Guerra file sup. 11.

[AGS Guerra, 572] General Archive of Simancas, Guerra file 572.

[AGS Guerra, 5759] General Archive of Simancas, Guerra file 5759.

[Actas, 1765-] Actas del Colegio Militar de Caballeros Cadetes del Real Cuerpo de Artillería desde el año 1765. Da principio en el año 1765 y concluye en fin del año 1787. Library of the Academy of Artillery, Segovia. (Minutes of the College).

[Calabro, 1724] Geometría Elemental, y Practica, en el suscinto methodo q. la enseñó en Barcelona el año de 1724 don Mateo Calabro ingeniero ordinario, i maestro de la escuela de artillería y náutica establecida en dicha capital. Madrid, Biblioteca Nacional de España. Course notes of Blas de Lana in 1724. Accessible in the web: `http://bdh-rd.bne.es/viewer.vm?id=0000140103\&page=1` (30/11/2021)

[Lucuce, 1759] P. Lucuce. Curso Matemático para la instrucción de los militares. Introduccion. [Tratado I Aritmetica]. Madrid, Biblioteca Central Militar. 1759.

[Lucuce, 1761] P. Lucuce. Tratado V De la artillería. 1761. Madrid Biblioteca Central Militar. Both are the course notes of Antonio Remón Zarco del Valle, when he was cadet. Accessible in the web: `https://bibliotecavirtual.defensa.gob.es/BVMDefensa/i18n/consulta/registro.cmd?id=4532` (30/11/2021)

[Martínez, 1756] Relación del estado actual de la real Escuela de Artillería de Cádiz y del curso que el Comisario ordinario Dn Gabriel Martínez ha compuesto para su establecimiento con las máximas que por su dirección se han practicado [AGS Guerra sup. 11].

[Silva, 1774] F. Silva. Primera parte del álgebra o Análisis Matemático para la instrucción y uso de los cavalleros cadetes del Real Cuerpo de Artillería que se educan en la Militar Academia de Segovia. Private collection, 1774. Course notes of Felipe Silva, taken when C. Vimercati was First Professor.

[Silby, 1754] J. R. Silby. Tratados que se darán y coordinación de ellos en la escuela de Matemáticas establecida por orden de S. M. de 21 de octubre de 1751 [AGS, Guerra 571].

Printed

[Anonymous, 1693] Anonymous (Attributed to José Chafrión). *Escuela de Palas o sea Curso Mathematico dividido en XI tratados. Marcos Antonio Pandulpho Malatesta.* Milan, 1693.

[Álvarez, 1734] G. Álvarez. *Conclusiones mathematicas, dedicadas al serenissimo Principe D. Phelipe ... Presididas por el R. P. Gaspar Álvarez.* Madrid. 1734.

[Benavente, 1762] M. Benavente. *Conclusiones de mathematicas ... Las preside el P. Miguel Benavente.* Joachin Ibarra, Madrid, 1762.

[Berenguer, 2020] J. Berenguer. Introducing differential calculus in Spain: The fluxion of the product and the quadrature of curves by Tomás Cerdá. *British Journal for the History of Mathematics*, 26: 23–49, 2021.

[Bramieri, 1760] E. Bramieri. *Conclusiones Mathematicas, defendidas en el Real Seminario de Nobles ... Presididas por el Padre Estevan Bramieri.* Joachin Ibarra, Madrid, 1760.

[Capel, 1988] H. Capel, J. E. Sánchez and O. Moncada. *De Palas a Minerva.* CSIC-Serbal, Barcelona, 1988.

[Cerda, 1758] T. Cerda. *Liciones de Matematica, ó Elementos Generales de Aritmética y Algebra para el uso de la clase ... Tomo segundo.* Francisco Suria, Barcelona, 1758.

[Cerda, 1764] T. Cerda. *Leccion de Artillería para el uso de la clase.* Francisco Suria, Barcelona, 1764.

[Certamen, 1754] *Certamen Académico Mathematico que obtuvo la Real Escuela del Cargo del Cuerpo General de Artillería establecida en la Plaza de Barcelona en los días 29. 30. y 31. de Octubre de 1754.* Joseph Altés, Barcelona.

[Escrutinio, 1754] *Descripción de los escrutinios quinto y sexto, celebrados en la Real Escuela de Mathematicas del cargo de el Cuerpo general de Artillería, en la Plaza de Cádiz, los días 24 y 25 de mayo de 1754.* [AGS, Guerra 572].

[de Medrano, 1680] S. Fernández de Medrano. *El practico artillero que contiene tres tratados. En el primero se declaran las piezas de cada genero y sus diferencias y se enseña el modo de cortar las cucharas y afustes, y el designio de las baterias. En el segundo se trata del manejo del cañon con algunas advertencias sobre sus alcances. Y en el tercero se trata en breve del puesto de cada official y otras advertencias.* Francisco Foppens, Brussels, 1680.

[de Medrano, 1688] S. Fernández de Medrano (Euclid). *Los seis primeros libros, onze, y doze, de los Elementos Geometricos del famoso philosopho Euclides Megarense.* Widow of Henrico Verdussen, Antwerp, 1688.

[Giannini, 1779] P. Giannini. *Curso Matematico para la enseñanza de los caballeros cadetes del Real Colegio Militar de Artilleria. ... Tomo I.* Joachin Ibarra, Madrid, 1779.

[Giannini, 1782] P. Giannini. *Curso Matemático para la enseñanza de los Ca-

balleros Cadetes del Real Colegio Militar de Artilleréda. . . . . Tomo II. Antonio Espinosa, Segovia, 1782.

[Giannini, 1795] P. Giannini. *Curso Matemático para la enseñanza de los Caballeros Cadetes del Real Colegio Militar de artillería. . . . Tomo III.* Antonio Espinosa, Segovia, 1795.

[Giannini, 1803] P. Giannini. *Curso Matemático para la enseñanza de los Caballeros Cadetes del Real Colegio Militar de Artillería. . . . Tomo IV.* Imprenta del Real Acuerdo y Chancillería, Valladolid, 1803.

[Herrero, 1990] M.D. Herrero Fernández de Quesada. *La enseñanza militar ilustrada El Real Colegio de Artillería de Segovia.* Academia de Artillería, Segovia, 1990.

[Massa-Esteve, Roca-Rosell, Puig-Pla, 2011] M.R. Massa-Esteve, A. Roca-Rosell, and C. Puig-Pla. Mixed mathematics in engineering education in Spain: Pedro Lucuce's course at the Barcelona Royal Military Academy of Mathematics in the eighteenth century. *Engineering Studies*, 3: 233–253, 2011.

[Morla, 1784] T. de Morla. *Tratado de artillería para el uso de la Academia de caballeros cadetes del Real Cuerpo de Artillería, . . . Tomo primero.* Antonio Espinosa, Segovia, 1784.

[Morla, 1785] T. de Morla. *Tratado de artillería para el uso de la Academia de caballeros cadetes del Real Cuerpo de Artillería . . . Tomo segundo.* Antonio Espinosa, Segovia, 1785.

[Morla, 1786] T. de Morla. *Tratado de artillería para el uso de la Academia de caballeros cadetes del Real Cuerpo de Artillería . . . Tomo Tercero.* Antonio Espinosa, Segovia, 1786.

[Navarro-Loidi, 2013] J. Navarro Loidi. *Don Pedro Giannini o las matemáticas de los artilleros del siglo XVIII. Asociación Cultural Ciencia y Artillería,* Segovia, 2013.

[Ordenanza, 1739] *Ordenanza, e instrucción para la enseñanza de las Mathematicas en la Real, y Militar Academia, que se ha establecido en Barcelona.* Antonio Marín, Madrid, 1739.

[Ordenanza, 1751] *Ordenanza, e instruccion, que se ha de observar en las Escuelas de Mathematicas, que con el titulo de Artilleria ha mandado el Rey erigir en las Plazas de Barcelona, y Cádiz bajo la dirección del Cuerpo General de ella.* Antonio Marín, Madrid, 1751.

[Ordenanza, 1768] *Ordenanzas de S. M. para el Real Colegio Militar de Caballeros Cadetes de Segovia.* Joachin Ibarra, Madrid, 1768.

[Reglamento, 1762] *Reglamento del nuevo pie en que Su Majestad manda se establezca el Real Cuerpo de Artillería.* Antonio Marín, Madrid, 1762.

[Terreros, 1744] E. Terreros. Conclusiones mathematicas, dedicadas al serenissimo y eminenísimo señor Don Luis de Borbon ...Presididas por el Reverendo padre Estevan de Terreros. Manuel Fernández, Madrid, 1744.

[Terreros, 1751] E. Terreros. *Conclusiones mathematicas, practicas, y especulativas, defendidas en el Real Seminario de Nobles ... bajo la instrucción y magisterio del Padre Estevan Terreros y Pando.* Imprenta del Supremo Consejo de la Inquisición, Madrid, 1751.

[Tosca, 1757] *Compendio Mathematico en que se contienen todas las materias mas principales de las Ciencias, que tratan de la Cantidad.* Imprenta de Joseph Garcia, Valencia, 1757, 9 v. First edition 1707–1715.

Mathematical Teachings in Italy in Military Schools in the Eighteenth Century and in the Age of Napoleon

Elisa Patergnani
University of Ferrara, Italy
elisa.patergnani@unife.it

Maria Giulia Lugaresi
University of Ferrara, Italy
mariagiulia.lugaresi@unife.it

Abstract

The history of mathematical teaching must consider the technical-scientific training organized and given in schools for the education of artillerymen and engineers in the eighteenth century, remodeled by Napoleon in the nineteenth century to bring it into line with the French approach. In these schools there were born the first study programs which taught how innovative mathematical results produced by thinkers like Galileo, Newton, and Euler could be applied to the field of war, as they were by thinkers like Tartaglia and others. Illustrious mathematicians taught there, such as Monge in Mézières and Lagrange in Turin. In Italy, the first model of artillery schools was represented by the Turin School which was established after the War of the Spanish Succession (1701-1714). Its mathematics study program was expanded with the teachings of young Lagrange on differential calculus. In the Napoleonic period eighteenth-century institutes were perfected in order to fit the Polytechnic model, as demonstrated by the school of Modena, which counted Paolo Ruffini among its teachers, and the schools of Pavia, where Antonio Collalto taught. In southern Italy, French influence on military schools can be found in the Royal Polytechnic and Military School of Naples. In the present work, we will provide an overview of the most relevant Italian Military Schools between the end of the 18th and the beginning of the 19th century by focusing on both professors and mathematical textbooks, written for the mathematical teaching in these schools.

1 Mathematics in military schools

In the early modern era, the growing importance of artillery and bastion architecture, owing to the devastating impact of heavy firearms on medieval fortifications, required the training of personnel possessing in-depth technical knowledge of such matters. In fact, traditional educational models of nobility appeared anachronistic in the face of the rapid development of new military techniques. These techniques required that skills be acquired through a properly organized and oriented curriculum – a curriculum that could not be fully provided by the traditional training courses of the ancient court schools for pages or in those academies where young people were trained only in the arts of chivalry (fencing, horse riding, etc.), all of which testified to an old way of understanding the art of war. The development of artillery made solid scientific education, of which mathematics was the backbone, indispensable. From the sixteenth century onwards, an attempt was made to compensate for the lack of military technical personnel conversant with the new knowledge by inserting engineers, architects, chemists and mathematicians into the militias. These were individuals who had no organic link with the military apparatuses in question and who bonded together in professional guilds and served the various lords only temporarily and for pay.[1]

It was precisely mathematicians who brought decisive scientific results to war-related matters. Nicolò Tartaglia was the first to try his hand at a mathematical study of the trajectories of projectiles launched by artillery, with the work *Nova scientia, cioè inventione novamente trovata utilissima per ciascuno speculativo mathematico bombardiero ed altri* (Venice, 1537).[2] For this reason he is considered the father of ballistics in the modern sense. In this book, the author described the

---

[1]The starting data of this work are taken from [Patergnani, 2020].

[2]A digital edition of this work of Tartaglia can be found in: `https://matematicaitaliana.sns.it/opere/26/`

scientific foundation of ballistics, the measurement of calibers, territorial detection and ballistic instruments, including the first firing table for the firearms, that is, tables that show how the range increases or decreases depending on each angle of elevation. Realizing that the principles of Aristotelian physics were not sufficient to adequately describe the detailed shape of the trajectory, he adopted, through observation, the hypothesis that this trajectory was an arc of a circle since he did not know Apollonius' treatise on conics. However, it was Cardano who specified the trajectory to be in fact the arc of a parabola: "Cum vero pila ad supremum recta pervenerit, non per circulu, nec recta rursum illico descendit, sed media quasi linea, quae parabolae ferme imitatur circumambientem lineam".[5] All this anticipated Galileo's and Newton's studies on the principle of inertia. Tartaglia was also one of the first mathematicians to deal with fortifications, as can be seen from the list drawn up in 1810 by the architect Luigi Marini (1778-1838),[4] in which the first work to be cited is *Quesiti et Inventioni diverse* (Venice, 1546) by the Brescia-born mathematician.

Galileo also addressed the topic of ballistics, both indirectly through his early work consisting of studying the centers of gravity of solids (*Theoremata circa centrum gravitate solidorum*, whose composition can be traced back to the years 1587 and 1588),[5] and directly through the invention and diffusion of the geometric-military compass, to which he devoted the work *Le operazioni del compasso geometrico militare* (1606). The latter was an instrument of Galileo's own design, based on the principle of proportional quantities, which had the function of facilitating the mathematical calculations necessary to adjust the firing trajectories of the artillery. In his work, *Discorsi e dimostrazioni matematiche intorno a due nuove scienze* (Leiden, 1638), Galileo came to assimilate, starting with the erroneous assumption of catenary and parabola, the entire tra-

---

[3]See [Cardano, 1550, p. 59]: Cardano knew Apollonius' work about conics: "Scripsit octo de conicis elementis libros egregios, quorum ad hanc usos diem primi tantum quatuor editi sunt, tam foede a translatore conspurcati, ut ne illos etiam aeditos merito dicere possis", Ivi, p. 314.

[4]This bibliographic list, which also contains information on the authors, is contained in [Marini, 1810], edition of the work of the military architect Francesco De Marchi (1504-1576) edited by Marini; on this see [Patergnani, 2020, pp. 50–54]. On the Marini see [Ravioli, 1858].

[5]See [Belle Sisana, 2022, p 492].

jectory to a parabola. However, Bonaventura Cavalieri (1598-1647) was one of the first to provide an exact demonstration of this in *Lo specchio ustorio* (1632). Aside from the definition of the trajectory curve, Galileo also realized that neglecting the problem of air resistance tends to lead to subsequent ballistic calculation errors. He also addressed the topic of military architecture by preparing two texts for students studying mathematics to undertake military courses. These latter were entitled *Breve instruzione all'architettura militare* and *Trattato di fortificazione*.[6]

Subsequently, Newton described the trajectory of a bullet as a curve formed by two branches of a parabola, of which the ascending branch is greater than the descending branch (*Principia*, 2nd ed. 1713). He also posed the problem of the influence of air resistance on the trajectory, thus deducing that the real trajectory must differ from the parabolic trajectory in a vacuum. According to Newton, therefore, to calculate the elements of the former, it was first necessary to know with some precision the effects of air resistance. However, even though he formulated the laws of air resistance, the impossibility of measuring some fundamental parameters (such as the speed at which the bullet emerged from the muzzle) prevented him from obtaining results consistent with his theoretical calculations. The very important scientific theories developed, and experiments carried out, from the sixteenth century onward sufficiently demonstrate the importance that mathematical studies have enjoyed (and still enjoy) in the art of war. Nevertheless, in the seventeenth century mathematical notions, even if they were part of the cultural formation of members of the military, remained for a long time rather an autodidactic matter than the subjects of any formal instruction. For example, much use was made of the personal private study of the great mathematical treatises, such as the *Cursus mathematicus* (Paris, 1634–1637) by Pierre Hérigone (1580–1643) and the *Cursus seu mundus mathematicus* (Lyon, 1690) by Claude François Milliet Dechales (1621–1678). To find the first institutionalized forms of mathematical teaching aimed at the military, we must look to the ecclesiastical colleges of Catholic Europe (especially the Jesuit ones), which were nevertheless intended for the training of the entire ruling class (including

---

[6]The treatise entitled *Fortificazioni* includes two parts, *Breve instruzione all'architettura militare* and *Trattato di fortificazione*. See Galilei (1891).

the military). In this context, an important role was played by the Roman College, a Jesuit university training institute, which had among its teachers Cristoforo Clavio (1538–1612), Cristoph Grienberger (1561–1636) and Athanasius Kircher (1602–1680). In the eighteenth century Ruggero Boscovich (1711–1787), the architect of the introduction into Italy of Newtonian physics, also taught there.

The first secular attempts to create schools dedicated exclusively to military education and technical training can be found in different countries. In the territory of the United Provinces, although no real military school was organized, the influence of the theories of Simon Stevin (1548-1620) was strong, thanks above all to the interest of Maurice of Orange-Nassau (1567-1625), for whom the mathematician had worked as a hydraulic engineer in the design and construction of dams and canals. Stevin, who had taught mathematics at the University of Leiden (organizing, among other things, the study of mathematics in the engineering school annexed to the latter), also carried out various studies on the determination of longitude and tides which were crucial contributions to the development of the Dutch sea fleet. His theories on fortification, expounded in the works *Sterckten-bouwingh* (1624), *Festung-Bawung* (1623), and *Nieuwe maniere van sterctebou, door spilsluysen* (1617), were collected among the *Oeuvres Mathematiques* published in French by Albert Girard in Leiden in 1634.[7] An attempt to found a centre for military training can be found in the Low Countries. In 1675 the Royal Military Academy of the Low Countries was established Brussels, thanks to the initiative of the Spanish military engineer Sebastián Fernández de Medrano (1646-1705). At this time, the region of the Low Countries belonged to the Spanish Crown. Medrano was the only director of the Academy of Brussels, that was closed in 1706.

Among the most outstanding students of the Academy of Brussels was Jorge Próspero de Verboom (1667-1744), who was in charge of the creation of the Spanish Corps of Military Engineers. In 1712 Verboom presented a project for a military academy based on the Royal Military Academy in Brussels. After the end of the War of Spanish Succession (1715), the king established the creation of an academy in Barcelona, which opened in October 1720. Verboom, who was in charge of the con-

---

[7][Marini, 1810, I, part II, p. 39].

struction of the Citadel and other military tasks, presented his idea of military education and proposed a detailed syllabus, focusing on the fact that a military officer should have a deep mathematical education. Verboom thought that the military engineers should have both theoretical and practical education, which included pure and mixed mathematics.[8] The first professor of mathematics and director of the academy was Mateo Calabro (from 1720 to 1738). Since 1737 Pedro Lucuce, who would be director from 1738 to 1779, proposed a new orientation for the studies in the Academy, that was reproduced in the Royal Ordinance of 1739. The document presented a detailed description of the academy and its courses.[9]

The numerous wars that shook Europe at the beginning of the eighteenth century gave decisive push towards the creation and istitutionalization of schools dedicated specifically to the technical-scientific education of the staff of military apparatuses. The following on one another's heels of the Spanish (1701–1714), Polish (1733–1738) and Austrian (1740-1748) Wars of Succession over the span of about sixty years, as well as the Seven Years War (1756–1763) - conflicts which involved all the major European states and also saw the Italian territory become the scene of important battles - created an urgent need to establish a corps of military engineers and to build schools for the training of technicians no longer borrowed from civil engineering, but rather forming an integral part of the army.[10]

In 1737, the publication of the notes on the new construction techniques for fortifications by Sébastien Le Prestre (1633-1707), Marquis of Vauban, paved the way for a growing awareness of this need on the

---

[8]As regards the 16th and 17th century in the Spain of the first Habsburgs the *Academia Real Mathematica* was created and remained active between 1582 and 1625. This was an institution dedicated to the education of future engineers and gunners. On the teaching in Spanish schools for the training of gunners and engineers see [Navarro Loidi, 2013a]; [Navarro Loidi, 2013b]; [Medina Ávila, 2014], [Velamazán and Ausejo, 2020], [Blanco and Puig-Pla, 2020], [Navarro Loidi, 2020]. On the teaching of mathematics in the Royal Academia of Barcelona see [Massa-Esteve et al., 2011] and [Massa-Esteve, 2014]. On Spain and European military schools see also [Patergnani, 2020, pp. 34-49].

[9]On the Royal Ordinance of 1739 and the course of Pedro de Lucuce see [Massa-Esteve et al., 2011, pp. 241–246].

[10][Verin, 1993].

part not only of European but also of Italian crowned heads.[11] There gradually developed an increasing interest in the training of troops and a renewed attention to the education of officers for what were later called "armi dotte" ("learned armies"). This consisted of an education, particularly in artillery and military engineering, which required in-depth knowledge of geometry and mechanics in regimental commanders, as well as knowledge of ballistics. In fact, the military disciplines imparted in the Colleges of the Nobles and in the Academies entrusted to the religious orders no longer guaranteed adequate training in these fields, and if the sons of noble families were still often consigned by their families to careers in the military, they were no longer in sufficient numbers to satisfy the demands of the new armies. Therefore, schools were created to professionally prepare the number of cadets necessary for the militia, with teaching programs focused on mathematical and physical disciplines. Thus in just less than thirty years, the most important European institutes dedicated to the training of officers were born, in particular the corps of engineers and artillery.

Already in 1720, Louis XIV founded five artillery schools, divided into theoretical and practical sections, intended for the military education of officers and aspiring gunners: Metz, Strasbourg, Grenoble, Perpignan (later transferred to Besançon) and La Fère. Each "school of theory" had a mathematics teacher; among the best known are Bernard Forest de Bélidor, Jean-Louis Lombard, Jean-Henri Hertenstein, Sylvestre-François Lacroix.[12] These institutions were to prove to be among the most enduring of the eighteenth century and, enjoying great European fame, were imitated by Austria, Spain and Italy.

One of the most important French schools was created in 1748, the *École royale du génie de Mézières*. This school was intended to provide military engineers with the technical skills necessary for the use of the new weapons of war, as well as with cartographic and strategic knowledge, which, together with solid physical-mathematical knowledge, appeared to be the best way to prevail in future wars. The courses

---

[11]On the influence of Vauban in Portugal see [Conde Massa-Esteve, 2818].

[12]On the French military schools of the 17th-18th centuries see [Le Puillon de Boblaye, 1858], [Chalmin, 1954]; T[Taton, 1986, pp. 511-615]; in particular on scientific teaching in French military and artillery schools in the eighteenth century see [Hahn, 1986]; [Alfonsi and Guilbaud, 2015].

thus included the study of arithmetic, geometry, mechanics, hydraulics, and experimental physics; practical teachings related to geometrical and perspective drawing, to plans for fortifications and to exercises of siege and defense of already-held positions were not neglected either. Gaspard Monge (1746-1818) was called to teach at the institute. During his assignment, he also addressed topics such as the cutting of stones and the problems inherent in moving masses of earth. The numerous observations made by Monge, and the experiments he carried out, on these subjects, which resulted in the development of rigorous geometric methods, formed the basis of the discipline which subsequently took the name of descriptive geometry. Thanks to him, the set of geometric knowledge and techniques related to the cutting of architectural blocks (so-called stereotomy) occupied an important place in the curriculum of the Mézières school. The practical needs of engineers, who often had to dig ditches or build defensive embankments with the consequent displacement of large quantities of earth, led Monge to grapple also with the practical problem of optimal earth displacement, now known as the Monge-Kantorovich problem. The question arose out of the need to decide where to transport each mound of earth from the excavation to the embankment under construction to minimize the work required. The problem, whose applications still have implications in various fields (such as fluid mechanics), was addressed by Monge by formulating a partial differential equation, which was later known as the Monge-Ampère equation.[13]

As mentioned, some of the first military schools in Europe followed the structure of the French organizational model. It was the case of the English experience, which, in 1741 saw the foundation at Woolwich (headquarters of the Artillery Regiment) of the *Royal Academy*, which had the aim of providing the members of the technical corps of the English army with the mathematical notions necessary for the proper performance of their service.[14] Although initially the teaching level of

---

[13] [Borgato and Pepe, 2020]. The topic is also of considerable importance today, as demonstrated, two hundred years after the death of the French mathematician, by the attribution of the Fields Medal to Alessio Figalli for his contributions to the theory of "optimal transport" and its applications to the theory of equations, partial differential differentials, geometric measurement theory and probability.

[14] [Bruneau, 2020].

this institute was not particularly high and was still very much inspired by the French experience of the artillery and engineering schools, which had proved to be avant-garde up to that time, things changed in 1742. In fact, that year the work *New Principles of Gunnery* by Benjamin Robins (1707-1751) was published. and was made famous above all thanks to the German translation that Euler made three years later with the addition of an extensive mathematical analysis.[15] Robins's was indeed the first analysis of the trajectories of bullets published with the indication of empirical values in respect of air resistance. Starting from Galileo's theories, the English author devised some measuring instruments (such as the ballistic pendulum and aerodynamic handling) suitable for calculating the speed of projectiles and measuring the resistance exerted by the air upon a projectile body; Robins' empirical use of it marked a turning point in the study of external ballistics. The extensive mathematical analysis, then carried out by Euler, did the rest, raising ballistics to a high level of maturity and sophistication as a science, as was demonstrated, shortly thereafter. by the creation of shooting tables suitable for the various types of firearms. The importance of these studies did not escape the careful magnifying glass of the French legislators, who immediately understood the scientific, but above all didactic, scope that they could potentially provide for the teaching imparted to students of artillery and "gunnery" schools.

The wars that had spread across Europe in the first half of the eighteenth century also involved Italy. Thus, a greater interest in training troops and, in particular, a new concern with the education and training of officers for the so-called "armi dotte" developed also in Italy. The Kingdom of Sardinia, the Kingdom of Naples, and the Republic of Venice all endeavoured to create schools of artillery and military engineering suitable for this purpose. In these institutes, skilled scholars explored the most innovative mathematical topics, contributing to the scientific development of the subject and creating real centers for the dissemination of mathematical thought. Later, the knowledge that was initially generated and concentrated in these schools also spilled over into more general technical education, starting with the university training of engineers. The history of mathematics teaching, therefore, cannot fail

---

[15][Steele, 1994]; [Bernett, 2009].

to take into due consideration the technical-scientific education given at these institutions.

2 The Study of Mathematics in the Schools of Turin Before and After Lagrange

One of the best examples of what was achieved in this field at this time was the *Regie scuole militari teoriche e pratiche di artiglieria e fortificazione* of Turin, which was established in 1739. In the first half of the eighteenth century, after the entry of the Kingdom of Sardinia into the European scene, this state set about instituting important enlightened reforms, which also involved the university and the army.

As regards academic education, the reforms had the primary purpose of forming a loyal governing and administrative class capable of supporting the sovereign in the process of modernizing the Kingdom, thus limiting the Jesuit monopoly on higher education; to this end, numerous chairs of the reformed University were entrusted to illustrious scientific personalities of the time. Among the professors whom Vittorio Amedeo II of Savoy called to be part of the new University of Turin in pursuance of his reform project (implemented between 1717 and 1721), a relevant role was played by the mathematician Ercole Corazzi, appointed by virtue of one the Piedmontese-Sardinian sovereign's *Regie patenti*, dated November 15, 1720.[16] Born in Bologna in 1669, Corazzi entered the benedictine order in 1689. His academic career began with some failures; in fact, he presented himself several times for the Chair of Mathematics of Padua University but always without success. Corazzi's article on the *Proposizioni della Quadratura del Cerchio*, printed in Chieti in 1705 and harshly criticized by Agostino Ariani in 1706, sealed his defeat as a scholar and marked the end of any attempt to renew his candidacy for the Chair of Mathematics in Padua.[17] In 1709 he was in

---

[16]See [Patergnani, 2017b].

[17]Agostino Ariani (1672-1748) was an Italian mathematician. He came from Naples and he studied mathematics and physics as an autodidact. He was "primary professor" of mathematics (*Primario Professore di matematica*) in the Studium of Naples, Nicolò Di Martino was one of his pupils. A biography of Ariani can be found in *Memorie della vita, e degli scritti di Agostino Ariani*, collected by Vincenzo Ariani and published in Naples in 1778.

Bologna and the following year he succeeded Vittorio Francesco Stancari (1678-1709) in the Chair of Algebra at the University of Bologna, thus surpassing the most famous mathematician in the Bolognese milieu, Gabriele Manfredi (1681-1761). He taught there for about a decade, and in 1711 was appointed Professor of Military Architecture at the new Institute of Sciences of the city of Bologna, which he himself had inaugurated with a solemn dedicatory address. Having become an abbot, he was called in 1720 to the University of Turin by Vittorio Amedeo to teach mathematics during Francesco D'Aguirre's reforms (1681-1748). For the Turin students Corazzi composed two unpublished manuscripts now preserved at the University Library of Turin.[18]

The Savoy reforms of the first half of the eighteenth century also strengthened the army. Above all, the intent was to consolidate the Duchy of Savoy's new role of power in the international framework. Inserted into the historical juncture of the Wars of Succession, the new kingdom took inspiration more and more clearly from the examples of the great absolutist states, particularly France and Prussia. Once the strategic importance of army's technical corps (artillery and engineers) was recognized, the Military Academy was reformed, a new military Arsenal was built (1739), and the new *Regie Scuole Militari teoriche e pratiche di Artiglieria e Fortificazione* were founded. The latter were used to train new cadres of military engineers. These schools were designed by Ignazio Bertola (1676-1755), "primo ingegnere del Re". They were conceived as military polytechnic schools, parallel and equivalent to the university, a place to train artillery and engineering officers who would later play key roles in the administration of the Savoy state.

The school course, spread over seven years, included five years of intense studies, in which, in the morning, the students attended lessons in arithmetic, algebra, plane geometry, trigonometry, geodesy, solid geometry, mechanics, and hydrostatics, while the afternoon was dedicated to military studies: design, fortification, mines, attack, and defense of strongholds. During the last two years, the students specialized in artillery (with exams on powders, firearms, and practical artillery) or in fortifications (drawing and tools).

---

[18]An appendix with Corazzi's printed works and manuscripts can be found in [Patergnani, 2017b].

The presence of military teachers of great stature, such as Gaspare Tignola (1710-1775), Ignazio Andrea Bozzolino (1719-1791), and Carlo Andrea Rana (1715-1804) also contributed to the progressive improvement of the scientific training imparted to the students, without forgetting the great contribution made by the lectures of civil professors of the fame of Giuseppe Luigi Lagrange (1736-1813) and Francesco Michelotti (1710-1787), the latter holding the post of Professor of Mathematics also at the city's university. These teachers also had the task of creating textbooks for use by students, textbooks which were then widely distributed in Italian and foreign military schools.[19]

Through the first texts used by the Turin students (manuscripts which the students themselves produced under the dictation of the official masters and substitutes for the masters), it is possible to trace the study programs envisaged in these first courses. From archive research, two manuscripts have emerged that preceded he key year of 1739 by a few years, the year in which the Turin artillery and fortification schools were founded. From the title page, it is clear that these texts were attributed to Lieutenant Vittorio Amedeo Conti, also known as Sinser. We know that he was the commander of the Savoyard artillery at the siege of Cuneo in 1744[20] and that he was part of the teaching staff of the new artillery schools founded by Ignazio Bertola.

The first text, *Corso d'Aritmetica*,[21] consists of 208 pages, the last

---

[19]The years Lagrange spent in Turin (1736-1766) represented a relevant period in his scientific career. He obtained some of his most important results regarding calculus of variations and the study of the equation of string vibration, that contributed to his scientific achievement and to the development of mathematics in Turin, recognised both nationally and internationally. Lagrange's departure from Turin in 1766 marked a turning point in the study of mathematics. Nor his fellow students Giuseppe Angelo Saluzzo and Gianfrancesco Cigna nor his pupil Daviet de Foncenex carried on advanced mathematical research. Mathematical studies in Turin in the second half of the 18th century had a more local and less original character and can be identified with the scientific works of Giambattista Beccaria (1716-1781) and Francesco Domenico Michelotti (1710-1787). See [Lugaresi, 2017]).

[20][Miscellanea, 1878, p. 535].

[21]Royal Library of Turin, ms. Saluzzo, 569: Corso d'/Aritmetica/Dettato nell'Accademia/d'Artiglieria/Dal Luogotenente V.A.C. Sinser/ Direttore della medema l'anno/1736 (c. III). Physical description: In folio (21,5 cm x 33 cm), total papers131, in tail 77 white papers. For a complete list of manuscripts and printed texts used in the Turin school, see [Patergnani, 2020]).

77 of which are left blank. It opens with a brief historical introduction, beginning with a synopsis of the best-known names involved in this discipline over the centuries:

Arithmetic is the most sublime and the most necessary subject both to civil life and to the acquisition and improvement of all other sciences; the glory of its invention is claimed to the Phoenicians, after whom the most famous philosophers, such as Pythagoras, Socrates, Plato, Aristotle, Nicomachus, Euclid, Apuleius, Vitruvius, Ptolemy, Archimedes, Pisano, Boetius, Alberto, Orontio, Delborgo, Tartaglia, Clavius, Galileo and many others, explained it. From their voluminous treatises, the clearest, shortest, and most intelligible and necessary rules were chosen.[22]

After providing the definition of arithmetic:

Arithmetic is a part of mathematics that has as its content discrete quantity, properties and quantities of numbers.[23]

The author introduces the concept of "number", illustrating the subdivision between simple numbers (those smaller than 10), "articoli" numbers (those divisible by 10, such as 10, 20, 30, etc.) and composite numbers (formed by simple numbers and "articoli' numbers, such as 11, 12, 17, 25, etc.).

Before turning to deal with the functioning of the numerical system used by the Romans, the author provides a brief historical note on the introduction of Arabic numerals into the Italian peninsula, stating that in the opinion of some authors this was likely to have been the work of the Goths.

---

[22]La più Sublime, et più necessaria tanto all'viver civile, che all'acquisto, et perfezione di tutte le Scienze sì è l'Aritmetica; la gloria della di cui Inventione si attribuisce a Fenici, doppo li quali l'Illustrarono li più famosi Filosofi, come Pittagora, Socrate, Platone, Aristotele, Nicomaco, Euclide, Appulejo, Vitruvio, Tolomeo, Archimede, Pisano, Boetio, Alberto, Orontio, Delborgo, Tartaglia, Clavio, Gallileo, et infiniti altri dalli di cui voluminosi trattati si sono scielte le più chiare, brevi, intelligibili, et necessarie regole, che qui jnfra si spiegheranno. Ivi, *Introduzione all'Aritmetica/Proemio*, cc. 1-2. The manuscript contains only the list of authors, without references to the titles of the works.

[23]"l'Aritmetica è una parte delle matematiche qual hà per oggetto la quantità discreta, proprietà, e quantità de numeri". Ivi.

He then enunciates the principles of arithmetic: addition, subtraction, multiplication, and division. The topics covered concern operations with whole numbers, the different numbering systems and coins, operations with "i rotti", proportions with the rules of three and false position, root extraction, ending with a final 18 pages devoted to the first operations conducted between algebraic quantities.

The second treatise (1737) *Geometria Pratica, Geodesia, Trasfigurazione de piani, uso degl'Instromenti Matematici, Trigonometria, e Meccanica*, consisting of 209 pages, the last 18 of which are left blank, deals with the main problems of geometry as applied to the maneuvers of artillery and ballistics, and with the use of mathematical instruments (level, square, graduated semicircle, proportional compass, etc.) depicted in the frame of the page containing the title. Part of the work expounds on the principles of trigonometric calculus and its applications (papers 66r-79v).

With the reorganization of the schools by Bertola, the topics covered in class remained substantially unchanged, as can be seen from the manuscripts, dating back to the years of his directorship (1739-1755) and relating to the Euclidean plane and solid geometry and, once again, to arithmetic. A decisive turning point in the study programs came only with the appointment of Lagrange by Alessandro Pacino D'Antoni (1714-1786), who had been called upon to direct the theoretical schools on Bertola's death.

In 1755 Lagrange, at the age of just 19, was appointed as a substitute for the Master of Mathematics in the Turin artillery schools, with the task of drafting texts for the use of these schools' students. In particular, he created a manuscript on differential and integral calculus titled *Principj di Analisi Sublime*, now preserved in the Royal Library of Turin. It is a fundamental text because it allows us to ascertain just which topics Lagrange decided to teach his students and to prove how advanced the contents of that teaching were for the time.

The manuscript is divided into two parts: the first is devoted to the algebraic theory of curves, and the second to differential and integral calculus. The first contains a treatise on conic sections and algebraic curves, in which the properties of the curves are derived from their equations. Several paragraphs are devoted by the author to the construction of equations, that is to the attempt to determine "two curves that can be built that have among the abscissas of their intersections the solutions of the given equation."[24] Thus far, the topics have been addressed without making use of infinitesimal calculus. The second part initially develops an algebraic calculus of finite differences, introducing, starting from the latter, the actual differential calculus. The author, though using Leibnizian notations, mainly bases himself on Newton's ideas, as is evidenced by a long discussion which anticipates the exposition of the differential calculus, devoted to the calculation of finite increases and presented "as calculus of ultimate ratios of vanishing quantities'.[25]

Geometrical applications to subtangents and maximums and minimums then follow, with the author subsequently providing an explanation of the comparison between infinitesimal calculus and differential calculus, demonstrating the equivalence of the results.[26] The manuscript then moves on to deal with definite integral calculation as a consequence of finite integration, taking an approach similar to the previous one. Although the scientific standard of the Lagrangian manuscript was very high, D'Antoni, while acknowledging its theoretical excellence, did not spare the author criticism regarding the alleged poor applicability of the notions contained therein. According to him, these enunciations risked being poorly applicable to the art of war and therefore not very useful in a practical sense for students who were preparing to embark upon a military career. The remarkable complexity of the work, therefore, perhaps constituted its limitation and caused it to become much less widely circulated among the students of the artillery school than was the text drawn up by D'Antoni himself (*Principj di matematica sublime*, Turin, 1779). In fact, from a comparison between the two volumes, it can be seen that many topics overlap: the study of algebraic equations, Cartesian

---

[24]"Due curve 'costruibili' che abbiano tra le ascisse delle loro intersezioni le radici dell'equazione data." [Borgato and Pepe, 1987, p. 35]

[25]"Come calcolo delle ultime ragioni delle quantità evanescenti' [Borgato and Pepe, 1987, p. 26].

[26]In his studies on the foundations of infinitesimal calculus, Lagrange gave great importance to the calculus of differences. In the treatise *Principj di Analisi Sublime*, before exposing differential calculus, Lagrange developed an algebraic calculus of finite differences. The differential calculus determines "the ultimate ratios of the difference dy/dx, i.e. the ultimate terms to which the general ratios of the differences continuously approach, while these continuously decrease".

geometry, differential calculus with the study of curves, integral calculus and calculation of logarithmic and exponential quantities. Furthermore, going into details, D'Antoni's work presents various uncertainties and a certain superficiality in the basic definitions (for example, the definition of function is not provided and the differential calculus is presented only in a way which involves a series of conceptual misunderstandings).[27]

Nevertheless, Lagrange's decision to introduce differential and integral calculus into the study path for gunners represented a real turning point: it constituted, in fact, an essential tool for the study of the problems of mechanics and geometry and therefore of artillery and fortifications. In the 1730s, in university cities where there were no military schools such as Pavia, Ferrara and Bologna, there took shape an advanced teaching of "sublime calculus" (a name that became current at the time as a joint designation of differential and integral calculus). Also in Padua there was a "private" teaching of calculus.[28]

The knowledge initially concentrated in the artillery and engineering schools in Italy then spilled over into wider technical education, starting with the university training of engineers and laying the foundations for the study programs of technical schools born as a result of the Industrial Revolution.

The school of artillery and fortification of the Kingdom of Sardinia laid the foundation for the subsequent organization of similar institutions in the Kingdom of Naples (1745) and the Republic of Venice (1759). As in Turin, also in Naples, thanks to the presence of mathematicians such as Nicolò Di Martino (1701-1769) and Vito Caravelli (1724-1800), the creation of these schools, which were suitable for training cadets for the militia in the skills necessary for their profession, involved the adoption of advanced teaching programs which included the study of differen-

---

[27][Borgato and Pepe, 1987, pp. 31-33]. According to D'Antoni, Lagrange's treatise was "too lofty, metaphysical and lacking in application to the professions of artillery and engineering". [Borgato and Pepe, 1987, p. 22].

[28]In 1707 Jacob Hermann (1678-1733), a favourite pupil of Jacob Bernoulli, was called in Padua to occupy the Chair of Mathematics, which had become vacant. Hermann taught the Leibnizian calculus in his private lessons, while he devoted his public lectures to more basic subjects, such as classical geometry, mechanics, optics, hydraulics, and gnomonics. On Hermann's teaching of mathematics in Padua, see [Massa-Esteve, 2014, pp. 27–102].

tial and integral calculus.[29] On the other hand, the *Militar Collegio* of Verona, despite being directed by Anton Maria Lorgna (1735-1796), a mathematician attentive to new analytical methods, still presented study programs which excluded the study of the methods of Leibniz and Newton and remained anchored to the old programs, based solely on Euclidean arithmetic and geometry. The Veronese cultural milieu was extremely influenced by what happened at the University of Padua with the mathematical teaching of Giovanni Poleni and by the presence in Verona of Giuseppe Torelli, strongly anchored to the Euclidean tradition.[30]

3 Mathematics in the Italian Schools for Gunners of the Napoleonic Period

The Napoleonic campaigns of the late eighteenth century in Italy overwhelmed the absolutist monarchical institutions of ancient Italian states, causing the wind of the ideals of the French Revolution to blow also on the Italian Peninsula. Indeed, the military campaign begun in 1796 by Napoleon Bonaparte caused an authentic geopolitical revolution in Italy, which culminated in the expulsion of the rulers of the *Ancien Regime* from their respective thrones and the formation of the so-called "sister republics", in which the management of public affairs was based on the model of French institutions, with a strong centralization of state functions and the creation of a powerful bureaucratic system. This inevitably also had a decisive influence on the military education models which had been employed up to that time, leading to the adoption of the new Napoleonic school system in Italy. In France, the starting point for those wishing to pursue a military career was the *École Polytechnique*, established in 1794 to train specialists in the civil and military technical services of the state. Attending this school was a prerequisite for subsequent access to schools applying scientific knowledge to artillery and military engineering, since it provided aspiring military personnel with basic knowledge of physics and mathematics. Therefore, the Napoleonic

---

[29] On the contents of the mathematical teachings in Naples see [Patergnani, 2020, pp. 107–123].

[30] About this topic, see [Patergnani, 2017a]; [Patergnani, 2020].

Empire inherited an excellent military education organization. Along-side the *École Polytechnique*, it could also count on regimental schools in which gunners and non-commissioned officers were trained. This proven system was improved by Napoleon through the establishment, first in France and later in the territories occupied by the *Grande Armée*, of a series of high schools in which mathematics came to play a fundamental role. The establishment of these schools in fact represented a milestone and model in the history of the teaching of mathematics, since it indicated that finally, after many centuries, this latter subject had achieved something like an equal status with Latin within the educational system.

The next step in the Napoleonic educational strategy was the establishment of schools reserved for the military, their children, and young scholars who wanted to pursue military careers. This process also affected Italy, which was divided into territories directly or indirectly linked to France (continental Italy was at this time divided into three large aggregations: the French Empire, which also included Piedmont, the Kingdom of Italy, and the Kingdom of Naples), in which military schools of the eighteenth century were perfected according to the French model.

In the territories which constituted the Cisalpine Republic, which later became the Italian Republic, and still later merged into the Kingdom of Italy (1805-1814), the recruitment of officers took place through the *Scuola Militare del Genio e dell'Artiglieria* of Modena directed by Leonardo Salimbeni (1752-1823), who was Lorgna's pupil and successor at the *Militar Collegio* of Verona, and in the military schools of Pavia.

The cultural and educational tradition of *Militar Collegio* was maintained by Napoleon with the establishment of a new military school in Modena. In the last years of the eighteenth century, Verona and its territory had been the subject of continuous war events. On June 1, 1796, in fact, the city was invaded by Napoleonic troops who engaged in several battles with the Austrians, who attempted to regain possession of it. On July 9, 1797, the Cisalpine Republic was proclaimed and, on October 17, the treaty of Campoformio was signed, by which the Venetian Republic was divided between Austria and France. Verona, initially ceded to the Austrians (January 21, 1798), became divided, through the provisions of the Treaty of Lunéville of February 9, 1801, between the

Austrians and the French, following the border established by the river Adige. The city's status as a continuously disputed territory would not, have allowed classes of the Veronese military school to be conducted in a regular manner. The Executive Directorate of the Cisalpine Republic then deliberated, on the 9th Thermidor of Year V of the French Revolutionary calendar (July 27, 1797), the constitution of the *Scuola Militare del Genio, e dell'Artiglieria* in Bologna. However, the location of the institute was established soon after in Modena, through the law of 24th Brumaire of Year VI (November 14, 1797).

After various vicissitudes connected with the succession of war events that affected Italian territory (and which led to various changes of headquarters), the school finally settled definitively in Modena in August 1801, resuming its regular operation on October 21 of that year, with Colonel Antonio Caccianino (1764-1838) as Director.

The law of November 22, 1803 divided the courses given at the school into two two-year periods, one theoretical and one applicative; the first was followed by all students collectively and the second was differentiated, with artillerymen following one applicative course and military engineers another. The mathematical disciplines were thus concentrated in the first two years: mathematical analysis, general theory of curve equations and differential calculus, descriptive geometry, elements of geodesy, mechanics, electricity, magnetism, heat, elements of chemistry, exact and approximate drawing were dealt with in the first year; sublime calculus, solid and descriptive geometry, mechanical and hydraulic applications, geodesy, chemistry and geodetic drawing were explored in the second.

Many distinguished scientists and scholars counted among those entrusted with the teaching of mathematical disciplines: Antonio Cagnoli (1743–1816) for "sublime calculus", Paolo Cassiani (1743-1806) for descriptive geometry and hydrodynamics, Giuseppe Tramontini (1768-1852) for drawing and architecture. Upon Cassiani's death (on February 3, 1806), Paolo Ruffini (1765–1822) was called upon to teach mechanics and hydrodynamics. Although access to the school became, in principle, possible directly after the completion of high school studies, the mathematical knowledge required was much greater than that required in Napoleonic high schools. For the preparation of the students for the entrance exams, then, a *Corso di matematica ad uso degli aspiranti della scuola d'artiglieria e genio* (Modena, Società Tipografia) was published from 1805 to 1808 in 5 volumes (I arithmetic, II geometry, III algebra of Ruffini, IV trigonometry of Cagnoli, V an Appendix to Algebra of Paolo Ruffini, a booklet of Giuseppe Tramontini about the method of the three coordinates and the Elements of spherical geometry, with the main rules of spherical trigonometry, of Carlo Benfereri).

That part of this *Corso*, characterized by the greatest originality, appears to have been the *Algebra* written by Ruffini. The author openly follows the method used by Clairaut in his *Éléments d'algèbre*, thus starting from actual concrete problems to arrive at theoretical expositions of a general nature:

> That is why, in the present Course, I have mixed frequent examples with rules and precepts, why, according to Clairaut's method, I have decided to infer the necessity of algebraic operations from these very examples, and why, from time to time, some remarks are joined both to the former and to the latter, in order to better illustrate their correctness, their nature, and their various properties.[31]

This choice was intended to serve didactic purposes: the intention, that is to say, was to allow the student to immediately enter the spirit of the subject, avoiding the usual difficulties involved in having to learn a long series of abstract precepts and operations without immediately evident implications for their application. According to Ruffini, this teaching method would allow the student to better learn the more complex parts of the subject, avoiding the most frequent errors in mathematical investigations.

The first part of the work is devoted to the study of the four fundamental operations, followed by a discussion of fractions, prime numbers, and the study of arithmetic, geometric, and harmonic proportions. Chapter VII, devoted to mathematical analysis, is also of great interest,

---

[31] Ecco il perchè alle regole ed ai precetti ho nel presente Corso misti degli esempj frequenti, il perchè ho procurato giusta il metodo di Clairaut di dedurre dagli esempj medesimi la necessità delle operazioni algebriche, ed il perchè sì a queste, che a quelli unisconsi di quando in quando delle riflessioni atte a far meglio conoscere la loro giustezza, l'indole loro, e i diversi accidenti. [Ruffini, 1807, p. II].

since this latter is here defined as "the art ... of solving problems by means of algebra", that is to say, it is considered not as a discipline but rather as a method.[32]

The *Corso* had been published with a dedication to the Minister of War of the Italian Republic, General Domenico Pino (1760-1826); Antonio Cagnoli presided over the edition, which also involved providing, together with Cassiani, an opinion on the level at which the work should be pitched, a question later decided by Colonel Caccianino. The work is presented as an affirmation of Italianness, expressed by the choice of contributions, which strove to distinguish itself from French models also predominant in the cultural sphere.

In 1813, with the Anglo-Austrian landing at the mouth of the Po (November 20), the school was transferred to Reggio Emilia and then to Parma, returning to its original location on December 4. On January 20, 1814, the Modenese headquarters were invaded by Neapolitan troops, and the institute was handed over to the Duke of Modena Francesco IV d'Este who, however, declared that he did not intend to keep it among his possessions. On May 18, it was moved to Cremona and, on June 1, the school passed under Austrian jurisdiction, the Austrians showing particular interest in keeping it operational. Archduke John, Director General of the fortifications and of the Austrian Academy of Military Engineering, in fact invited Caccianino to present him with a project for the new organization of the institute. Nonetheless, at the end of July 1815 the school was closed by the decree of the Austrian commander, and on September 16 the dissolution report was drawn up.

During his military campaigns on Italian territory, Napoleon also invested heavily in another city in northern Italy, Pavia, which quickly became the main headquarters of the artillery of the Kingdom of Italy. The main reasons for this choice are to be sought not only in the location of the city, whose crossing by the river Ticino facilitated the transport of materials, but also in the fact that Pavia remained the seat of a university. The first step in this direction was the construction of an arsenal between 1803 and 1805, with the establishment in the city of an artillery regiment. On July 22, 1803 the *Scuola per il Reggimento*

---

[32]"l'arte ... di sciogliere i problemi col mezzo dell'algebra'. Ivi, p. 138. On the diffusion of the calculus in Italy see [Pepe, 1981],[Dupont Roero, 1991].

d'Artiglieria a Piedi was established through a special decree.

The Napoleonic decree prescribed admission requirements, which required the passing of a mathematics exam, and set the start of the courses for 15th September. Antonio Collalto (1765-1820) was appointed Professor of Mathematics and was to keep this post until his transfer to Padua, which took place in 1807. (Here, he held the chair of "Introduction to Sublime Calculus" at the University of Padua).[33]

Collalto had the merit of establishing for the school a high scientific standard. Once it had been ascertained just what level of scientific education the students had entered the school with, they were divided into two classes: one made up of less educated officers and non-commissioned officers (in which geometry, trigonometry, and the first elements were studied); the other comprised the more highly trained officers (this one devoted to the study of conic sections and "sublime calculus").

Collalto was also responsible for the first study texts for the students of this school. The mathematician, who after Campoformio had been forced into exile in Paris for his adhesion to the Jacobin movement and whose return to Italy had been made possible only by the Napoleonic victories in this country, developed two manuals for the Pavia schools: *Dell'istruzione teorico-pratica degli ingegneri* (Pavia, 1804) and *Geometria analitica a due coordinate* (Milan, 1806). The latter, in particular, was widely disseminated and served, above and beyond its narrower pedagogic purpose, also to describe Collalto's attempts to introduce the results of the most recent foreign studies on calculus in Italy, an aim he had already pursued a few years earlier in his most important theoretical paper: *Identità del calcolo differenziale con quello delle serie ovvero il metodo degli infinitamente piccoli di Leibnizio spiegato e dimostrato colla teoria delle funzioni di Lagrange* (Milan, 1802).

In the preface of *Geometria analitica*, the author provides a historical overview of the most significant theories on this subject starting from Descartes' revolutionary idea of applying algebra to geometry. He then complained about the difficulties many authors appeared to have in abandoning the previous approaches with which geometry was treated in ancient times, to the point of celebrating Lagrange and Monge, the first, according to Collalto, to have shown that all geometrical questions, even

---

[33]See [Dupont Roero, 1991].

those relating to the straight line and the circle, could only be handled through combinations of equations.

Without forgetting the didactic purposes that he wanted to pursue, Collalto recalled that the first author to have discussed the application of algebra to geometry in an elementary way was Lacroix – although this latter, he adds, had not yet succeeded in freeing himself from the use of analytical methods alone.

Convinced that it was essential to introduce students to new methods, while following by his own admission the model of Lagrange and Monge and of treatise writers such as Biot and Lacroix, Collalto tried to present the proofs and solutions in a more general and uniform way, devoting much space to exercises "because only with practice can the methods be applied easily."[34]

The intent was to definitively detach and separate the teaching of geometry from traditional intuitive models. In accordance with this approach, the lessons included first and second order lines, which justified the denomination of "two-coordinate" geometry, since they were lines in a single plane, "their points always lead back to two coordinates given by position in the same plane."[35]

His intention was to publish other lessons at some later point (which he wanted to call "three-coordinate" geometry lessons), with the declared hope of being the first in Italy to disseminate these new methods. This intention was realized when, on leaving Pavia and assuming the chair of Introduction to Sublime Calculus at the University of Padua in 1806, he published the *Nuove lezioni di geometria analitica a tre coordinate* (Padua 1809).

In Pavia there was another military school intended for training infantry officers and specialists in cavalry weapons. The institute, founded by an imperial decree of July 7, 1805, was part of the Napoleonic strategy of creating in the Kingdom of Italy a school similar to the *École spéciale militaire* of Fontainebleau, dedicated to the training of French officers in mêlée weapons (infantry and cavalry).

The first Professor of Mathematics was Collalto himself, replaced

---

[34]"poiché unicamente con questo esercizio si può ottenere la facilità delle applicazioni dei metodi."[Collalto, 1806, p. XII]

[35]"i loro punti si riportano sempre a due coordinate date di posizione nello stesso piano."

in 1806 by Paolo Tognola and a certain Gratognini, probably Giovanni Gratognini (1757-1824), pupils of Lorenzo Mascheroni, and in 1807 by the young Antonio Maria Bordoni.

To be admitted, candidates had to be aged between 16 and 20, and, in addition to being able to speak and write correctly in Italian, also had to demonstrate knowledge of arithmetic and geometry. The course lasted two years, and each year was divided into two semesters, theoretical and practical, starting respectively on 1st October and 1st April. Regarding the study texts, it was planned to translate those used in French high schools; in particular the engineer Carlo Paganini (1782-1844),[36] commissioned by the Ministry of War to edit an Italian edition of Bezout's elementary work, insisted adopting the complete course taught at the French school. His proposal took shape in 1810, when the minister ordered to print a compendium of the mathematics lessons of the military school of Saint-Cyr, entitled *Corso di matematica ad uso delle scuole militari del Regno d'Italia*, divided into two volumes (Arithmetic and Algebra; Geometry and Mechanics).

In this work, arithmetic was expounded only to the extent that was required to provide students with the necessary knowledge for the calculations relating to the administration and accounting of military units. In the discussion, algebra preceded geometry "being an obvious thing not to separate algebraic calculus from the numerical calculus, and because, on the other hand, Geometry derives much more benefit from Algebra than Algebra from Geometry."[37] The treatment of the latter, departed as little as possible from the discussions of Legendre and Lacroix. For obvious didactic purposes, the rigor of many demonstrations was limited, although efforts were always made not to damage the clarity of reasoning. The material on geometry was divided into six books; the first expounded the properties of lines and surfaces, solved some related problems and provided graphic solutions. The second book expounded the properties of planes and solids, including those of rotation. The measuring methods of some bodies that made up various fortification works were also discussed, a matter that was not dealt with in sufficient

---

[36][Giornale, 1845, pp. 39-40].

[37]"Essendo cosa naturale di non disgiungere il calcolo algebrico dal numerico, e perché d'altronde la Geometrìa molto maggiore vantaggio ricava dall'Algebra, che non ne trae l'Algebra dalla Geometrìa."[Corso, 1810, vol. I, p. 8].

detail by Bezout. Descriptive geometry was the subject of the third book, devoted in particular to the theory of projections, but limited to the parts of this theory that were indispensable to the students of imperial military schools who needed to follow the course on fortifications. The fourth book contained the theory and practice of levelling "because an officer may be charged with a more difficult action than drawing a small outline of entrenchment, this section has also been dealt with in greater detail than what Bezout has made."[38] In the fifth, the principles of rectilinear trigonometry and its application to the measurements of distances and heights were succinctly expounded; the geometrical processes that can be used to copy or reduce topographic maps were then briefly described. The elementary notions of analytic geometry found their place in the sixth book, described in such a way as to allow an understanding of the demonstration of some propositions of mechanics, to which the last part of the work was dedicated, with an examination of the application of the principles of statics to simple machines and other uses in artillery, ending with the exposition of dynamics and hydrostatics and with their natural application to the theory of projectile motion and the study of the barometer.

The work was closed with a table of definitions and principles useful for providing students with a synthetic picture of the mathematical theories expounded and with a sort of program that, on the one hand, would help them summarize the lessons of the professors and, on the other hand, make them aware of the level of relevant education that they had acquired through their own private study.

Despite its original intentions, the Pavia military school struggled, in the early years of its existence, to align itself with the study programs of the Fontainebleau school, since the minimum admission requirements and the level of education of the courses were lowered to facilitate enrollment requests.

The Napoleonic influence on the models of education and training adopted by Italian military schools is also evident in the Kingdom of Naples. On August 13, 1811 in this latter kingdom the *Scuola Reale Po-*

---

[38] "siccome può accadere che un ufficiale venga incaricato d'un'operazione più difficile di quella che consiste nel disegnare un piccolo profilo di trincieramento, così anche questa parte è stata trattata più minutamente di quello che Bezout non ha fatto." Ivi, p. 9.

litecnica e Militare was established similar to the French one. It aimed to spread the culture of the mathematical and chemical sciences, military art, drawing, and fine literature in southern Italy, as well as to train the students of schools specializing in applied military arts and disciplines, land and sea artillery, and military engineering. The following courses were held at this school: synthetic geometry, solids, trigonometry, and conic sections; analysis and its application to geometry; descriptive geometry, mechanics, and physics, principles of astronomy related to geography, military sciences, and the drawing of geographical maps; chemistry and its application to the arts, fine literature, geography and history, civil and military architecture, drawing (figures and landscape) and the French language.

As regards textbooks for use by students, these years saw the publication of the *Istituzioni di geografia fisica e politica*, in two volumes, by Luigi Galanti and Nicola Morelli, and the *Elementi di matematica composti per uso della Reale Accademia Militare* by Vito Caravelli, printed by Gennaro Reale, who was to publish the ninth edition, corrected and revised, in 1815.[39]

Due to the breadth of the curriculum, a specific series was created which collected together all the texts for the study of the scientific disciplines taught at the school, and which was entitled *Saggio di un corso di matematica per uso della Reale Scuola Politecnica, e Militare*, divided into thirteen volumes relating to: *Aritmetica* (I) and *Algebra* (II) by Giovanni Rodriguez, *Planometria* (*Geometria piana,* III), *Analisi a due coordinate* (V) and *Planometria* (*Trigonometria piana,* VI) by Ferdinando De Luca, *Stereometria o sia Geometria solida* (IV) and *Ristretto di Geometria descrittiva* (IX) by Gaetano Alfaro, *Analisi applicata alle tre dimensioni* (VII) and *Calcolo differenziale ed integrale* (VIII) by Ottavio Colecchi, *Meccanica* (X) and *Idrodinamica* (XI) by Nicola Massa, *Trigonometria sferica* (XII) and *Geografica matematica* (XIII) by Tommaso Farias. These texts were published between 1813 and 1815 at the Sangiacomo Printing House belonging to the same school. Bibliographic information regarding the authors of the course, and regarding

---

[39][Trombetta, 2011, p. 198]. The first edition of the treatise *Elementi di matematica composti per uso della Reale Accademia Militare dal professore di fisica sperimentale, e chimica, e direttore delle scienze della medesima Vito Caravelli* was published by the editor Raimondi in Naples in nine volumes in 1770-72.

the teachers at the polytechnic school, is scarce; the few references which exist can be found in [1924]: Rodriguez was born in Gaeta, De Luca in Serracapriola, Alfaro in Pescara, Massa in Genoa and Farias in Naples.[40] Missing from this list is Ottavio Colecchi (1773-1847), also a professor and author of manuals of the School, who, together with Ferdinando De Luca (1802-1885), was part of the group of so-called "analytics", that is, one of the two groups into which the Neapolitan mathematical school was divided at that time.[41]

4   Conclusions

The history of mathematical teachings cannot ignore the military schools created during the 18th century in the context of the great European wars. Over the centuries, mathematical teaching has gained more and more importance in technical-scientific training. In this work we focused our attention on the case of mathematics at the service of the art of war. The first ballistic studies are due to the Italian mathematician Nicolò Tartaglia who established an important tradition of studies and treatises on this subject, which was carried on in Italy by the Galilean school (Galilei and Cavalieri). Despite the importance of these studies, we have underlined that, regarding the art of war, mathematical notions did not yet have formal institutionalization in the 17th century military curriculum. At the European level, the first attempt to create a school entirely devoted to the education of the military was in Spain with the foundation of the *Academia Real Mathematica* (1582) and then with the creation of the Royal Military Academy of Mathematics of Barcelona (1720).

---

[40][Amodeo, 1924, p. 173].

[41]At the time, in fact, the Neapolitan scientific milieu more or less subsisted on the confrontation between, on the one hand, the group of supporters of synthetic methods (led first by Nicola Fergola and later by Vincenzo Flauti), who were mostly conservatives both in the sphere of mathematical research and in political matters since they openly supported the Bourbon government, and, on the other hand, the supporters of analytical methods, who were, by contrast, progressive on both fronts. For more on the different aspects of the contrast between synthetics and analytics, see: [Ferraro et al., 1995]; [Palladino, 1999]; [Gatto, 2000]; [Mazzotti, 2002]; [Ferraro, 2013].

The Wars of Successions that crossed Europe in the first half of the 18th century provided a valid reason to improve (if existing) or create military schools for the training of troops. We gave a wide overview of the Italian experience, with particular attention to the foundation of the *Regie Scuole Militari teoriche e pratiche di Artiglieria e Fortificazione* in Turin. Other case studies were those of Pavia (*Scuola per il Reggimento d'Artiglieria a Piedi*), Verona (*Militar Collegio*), Modena (*Scuola Militare del Genio e dell'Artiglieria*) and Naples (*Scuola Reale Politecnica e Militare*).

This article provides an overview of the syllabuses in these schools, remodelled in the Napoleonic era on the basis of French and Spanish models. The analysis of syllabuses, manuscripts of the courses, handbooks, and the teaching activities imparted by the teachers who taught in those schools (Joseph Louis Lagrange in Turin, Anton Maria Lorgna in Verona and so on) brings out an important chapter in the history of this discipline in Italy, anticipating a didactic model that would be applied in the 19th century technical education institutes.

References

[Alfonsi and Guilbaud, 2015] L. Alfonsi and A. Guilbaud. La guerre de Sept Ans (1756-1763) et ses conséquences pour les écoles militaires françaises. In C. Gilain and A. Guilbaud, editors, *Sciences mathématiques 1750–1850, Continuité et ruptures*, pages 127–155. CNRS Editions, Paris, 2015

[Amodeo, 1924] F. Amodeo. *Vita matematica napoletana: Studio storico, biografico, bibliografico.* Tipografia dell'Accademia Pontaniana, Napoli, 1924.

[Aquila imperiale, 1994] *All'ombra dell'Aquila imperiale. Trasformazioni e continuità istituzionali nei territori sabaudi in età napoleonica (1802–1814).* 2 voll., Ministero per i Beni culturali e ambientali, Roma, 1924.

[Barbieri and Cattelani, 2003] F. Barbieri and F. Cattelani Degani. I matematici italiani nel periodo napoleonico: i contributi di P. Cassiani, G. Tramontini, P. Ruffini alla scuola d'artiglieria e genio di Modena. In S. Lenzi, editor, *Il sogno di libertà e di progresso in Emilia negli anni 1796–97*, pages 120–126. Omicron, Modena, 2003.

[Belle Sisana, 2022] R. Bellé and B. Sisana. Galileo Galilei and the centers of gravity of solids: a reconstruction based on a newly discovered version of the conical frustum contained in manuscript UCLA 170/624. *Archive for*

History of Exact Sciences, 76: 471–511, 2022. https://doi.org/10.1007/s00407-022-00289-4.

[Berkel, 2005] K. van Berkel. Simon Stevin et la fondation de l'ecole militaire de Leyde en 1600. In C. Secretan, P. Boer Den, editors, *Simon Stevin. De la vie civile*, 1590, pages 93-108. ENS Éditions, Lyon, 2005.

[Binaghi, 2017] R. Binaghi. The teaching of mathematics, architecture and engineering. In K. Bjarnadottir, F. Furinghetti, M. Menghini, J. Prytz and G. Schubring, editors, *Dig where you stand 4. Proceedings of the 4 ICHME*, pages 31-46. Edizioni Nuova Cultura, Roma, 2017.

[Blanco and Bruneau, 2020] M. Blanco and O. Bruneau, editors. Les mathématiques dans les écoles militaires (XVIIIe-XIXe siècles). *Philosophia Scientiae*, 24/1, 2020.

[Blanco and Puig-Pla, 2020] M. Blanco and C. Puig-Pla. The Role of Mathematics in Spanish Military Education in the 1750's: Two Transient Cases. In [Blanco and Bruneau, 2020, pp. 81–95].

[Borgato and Pepe, 1987] M.T. Borgato and L. Pepe. Lagrange a Torino (1750-1759) e le sue lezioni inedite nelle R. Scuole di Artiglieria. Bollettino di Storia delle Scienze Matematiche, 7/2, 1987.

[[Borgato and Pepe, 1990] M T. Borgato and L. Pepe. *Lagrange. Appunti per una biografia scientifica*, La Rosa Editrice Torino, 1990.

[Borgato and Pepe, 2020] M.T. Borgato and L. Pepe. Gaspard Monge (1746-1818): application of geometry to analysis. *Eurasian mathematical journal*, 11/1: 13-24, 2020.

[Bottani, 2014] C.E. Bottani. I principi della dinamica di Tartaglia e Cardano nelle Cinquecentine dell'Istituto Lombardo. *Istituto Lombardo-Rendiconti. Classe di Scienze Matematiche e Naturali*, 148:131-142, 2014.

[Brizzi, 1976] G.P. Brizzi. *La formazione della classe dirigente nel Sei-Settecento: i seminaria nobilium nell'Italia centro-settentrionale*, Il Mulino, Bologna, 1976.

[Brizzi, 1992] G.P. Brizzi. Ritterakademien e seminaria nobelium. In G.P. Brizzi and J. Verger, editors, *Le università dell'Europa. III: Dal rinnovamento scientifico all' età dei Lumi*, pages 109-125. Silvana editoriale, Cinisello Balsamo, 1992.

[Bruneau, 2020] O. Bruneau. The Teaching of Mathematics at the Royal Military Academy: Evolution in Continuity. In [Blanco and Bruneau, 2020, pages 115-133].

[Cardano, 1550] G. Cardano. *De Subtilitate Libri XXI*, Norimbergae, apud Ioh. Petreium, 1550.

[Chiavolini and Roero, 2002] S. Chiavolini and C.S. Roero. Giambattista Beccaria, scienziato illuminista nel Regno sabaudo. In M. C. Vera de Flachs, editor, *Universidad e ilustración en America. Nuevas Perspectivas*, pages 21-41. Argentina, Hugo Báez Editorial, Cordoba, 2002.

[Chalmin, 1954] P. Chalmin. Les écoles militaires françaises jusqu'en 1914. *Revue historique de l'armée*, 10/2, 1954.

[Caruso, 1998] A.G. Caruso. Compendio di storia del pensiero balistico. *Quaderni di Oplologia*, 6: 35-60, 1998.

[Collalto, 1806] A. Collalto. *Lezioni Di Geometria Analitica A Due Coordinate*, Dalla Tipografia di Gio. Gius. Destefanis, Milano, 1806.

[Conrads, 1982] N. Conrads. *Ritterakademien der frühen Neuzeit. Bildung als Standesprivileg in 16. und 17. Jahrhundert*, Vandenhoeck & Ruprecht, Göttingen, 1982.

[Corso, 1810] *Corso di matematica ad uso delle scuole militari del Regno d'Italia tradotto dal francese per ordine di s.e. il sig. conte Ministro della guerra in seguito al decreto di s.a.i. il principe vice-ré*, 2 voll., Dalla Tipografia militare di Giuseppe Borsani, Milano, 1810.

[Dijksterhuis, 1970] E.J. Dijksterhuis. *The Principal Works of Simon Stevin*, C.V. Swets & Zeitlinger, Amsterdam, 1970.

[Dhombres, 1985] J. Dhombres. French Mathematical Textbooks from Bézout to Cauchy. *Historia Scientiarum*, 28: 91-137, 1985.

[Dupont Roero, 1991] P. Dupont and C.S. Roero. *Leibniz 84. Il decollo enigmatico del calcolo differenziale*, Mediterranean Press, Cosenza, 1991.

[Duveen and Hahn, 1957] D.I. Duveen and R. Hahn. Laplace's Succession to Bézout's Post of Examinateur des Elèves de l'Artillerie. *Isis*, 48/4: 416-427, 1957.

[Ferraro, 2013] G. Ferraro, Non sempre gli uomini che dimenticano hanno torto. Note critiche sulla storia della matematica nei territori napoletani. In G. Ferraro, editor, *Aspetti della matematica napoletana tra Ottocento e Novecento*, pages 9-139., Aracne, Roma, 2013.

[Ferraro et al., 1995] G. Ferraro, P. De Lucia and F. Palladino Franco. Alcuni tratti della matematica napoletana da prima a dopo la repubblica partenopea del 1799. *Rendiconto dell'Accademia di Scienze Matematiche e Fisiche*, 62: 225-274, 1995.

[Conde Massa-Esteve, 2818] A. Fialho Conde and M.R. Massa-Esteve. Teaching Engineers in the Seventeenth Century: European Influences in Portugal. *Engineering Studies*, 10/2-3: 115-132, 2018. https://doi.org/10.1080/19378629.2018.1480627.

[Fiocca, 1992] A. Fiocca. La geometria descrittiva in Italia (1798-1838). *Bollettino di storia delle scienze matematiche*, 12/2: 187-249, 1992.

[Gatto, 2000] R. Gatto. La discussione sul metodo e la sfida di Vincenzo Flauti ai matematici del Regno di Napoli. *Rendiconto dell'Accademia delle Scienze Fisiche e Matematiche*, 67, pp. 181-233, 2000.

[Giornale, 1845] *Giornale. compilata da varj dotti nazionali e stranieri*, tomo X, Presso la direzione del giornale, Milano, 1845.

[Karp and Schubring, 2014] A. Karp and G. Schubring. Mathematics Education in Europe in the Premodern Times. In A. Karp and G. Schubring, editors, *Handbook on the History of Mathematics Education*, pp. 129-151. Springer-Verlag, 2014.

[Hahn, 1986] R. Hahn. L'enseignement scientifique aux école militaires et d'artillerie. In Taton (1986), pp. 513–545.

[Bernett, 2009] H.J. Barnett. Mathematics goes ballistic: Benjamin Robins, Leonhard Euler, and the mathematical education of military engineers. *BSHM Bulletin*, 24: 92-104.

[Ilari, 2021] V. Ilari. Scrittori Militari Italiani dell'età moderna. *Dizionario bio-bibliografico 1419-1799*, Collana SISM, 2021.

[Le Puillon de Boblaye, 1858] T. Le Puillon de Boblaye. *Esquisse historique sur les ècoles d'artillerie*, Rousseau-Pallez, Metz and Tenera, Paris, 1858.

[Lorgna, 1996] Anton Maria Lorgna scienziato ed accademico del XVIII secolo tra conservazione e novità, *Accademia Nazionale delle Scienze detta dei XL*, Tipografia della Pace, Roma-Verona, 1996.

[Lugaresi, 2017] M.G. Lugaresi. *Vita scientifica di Giorgio Bidone. Torino dopo Lagrange*. Quaderni della Fondazione Filippo Burzio, Collana Studi e Ricerche, Torino, 2017.

[Marini, 1810] L. Marini. *Architettura militare di Francesco De'Marchi*, 8 voll., Da' Torchi di Mariano de Romanis e figli, Roma, 1810.

[Massa-Esteve et al., 2011] M. R. Massa-Esteve, A. Roca-Rosell, C. Puig-Pla. 'Mixed' mathematics in engineering education in Spain: Pedro Lucuce's course at the Barcelona Royal Military Academy of Mathematics in the eighteenth century. *Engineering Studies*, 3/3: 233-253, 2011.

[Massa-Esteve, 2014] M. R. Massa-Esteve. La Reial Acadèmia de Matemàtiques de Barcelona (1720-1803). Matemàtiques per a enginyers. *Quaderns d'Història de l'Enginyeria*. 14: 17-34, 2014.

[Mazzone Roero, 1997] S. Mazzone and C.S. Roero. *Jacob Hermann and the diffusion of the Leibnizian calculus in Italy*, Leo S. Olschki, Firenze, 1997.

[Mazzotti, 2002] M. Mazzotti. Le savoir de l'ingénieur. Mathématiques et politique à Naples sous les Bourbons. *Actes de la recherche en sciences sociales*, 141–142: 86–97, 2002.

[Medina Ávila, 2014] C.J. Medina Ávila. De la Escuela a la Academia. Los centros de formación de artilleros. *Revista de Historia Militar*, 1: 13-72, 2014.

[Mellado Romero, 2022] A. Mellado Romero. *La influencia del Cursus Mathematicus de Herigone en la algebrización de la matemática*. Phd thesis, Universidad de Murcia, 2022.

[Navarro Loidi, 2013a] J. Navarro Loidi. Don Pedro Giannini o las matemáticas de los artilleros del siglo XVIII, *Asociación Cultural "Biblioteca de Ciencia y Artillería*, Segovia, 2013.

[Navarro Loidi, 2013b] J. Navarro Loidi. La incorporación del cálculo diferencial e integral al Colegio de artillería de Segovia. *Llull*, 36/78, 2013.

[Navarro Loidi, 2020] J. Navarro Loidi. Foreign influence and the mathematics education at the Spanish College of Artillery (1764-1842). In [Blanco and Bruneau, 2020, pages 96-114].

[Palladino, 1987] F. Palladino. La matematica a Napoli nel Seicento e i suoi rapporti con l'Italia e l'Europa. *Giornale critico della filosofia italiana*, 66/3: 141-166, 1987.

[Palladino, 1999] F. Palladino. *Metodi matematici e ordine politico: Lauberg, Giordano, Fergola, Colecchi. Il dibattito scientifico a Napoli tra Illuminismo rivoluzione e reazione*, Jovene, Napoli, 1999.

[Patergnani, 2017a] E. Patergnani. The Teaching of Mathematics in the Italian Artillery Schools in the Eighteenth Century. In K. Bjarnadottir, F. Furinghetti, M. Menghini, J. Prytz and G. Schubring, editors, *Dig where you stand 4. Proceedings of the 4 ICHME*, pages 247-262. Edizioni Nuova Cultura, Roma, 2017.

[Patergnani, 2017b] E. Patergnani. Ercole Corazzi tra le Università di Padova, Bologna e Torino. *Bollettino di storia delle scienze matematiche*, 37/2: 267-297, 2017.

[Patergnani, 2020] E. Patergnani. *Gli insegnamenti matematici nelle scuole militari in Italia da Eugenio di Savoia a Napoleone*, Il Mulino, Bologna, 2020.

[Pepe, 1981] L. Pepe. Il calcolo infinitesimale in Italia agli inizi del secolo XVIII. *Bollettino di storia delle scienze matematiche*, 1/2: 43-101, 1981.

[Pepe, 2015] L. Pepe, Tra Università e Scuole Militari. Gli inizi degli insegnamenti del calcolo infinitesimale in Italia. In A. Romano, editor, *Dalla lectura all'e-learning*, pages169-181. Clueb, Bologna, 2015.

[Pepe, 2016] L. Pepe. *Insegnare matematica. Storia degli insegnamenti matematici in Italia*. Clueb, Bologna, 2016.

[Piva, 1996] F. Piva. Anton Maria Lorgna e la Biblioteca del Collegio Militare di Verona. In [Lorgna, 1996, pages 195-213].

[Ravioli, 1858] C. Ravioli. *Della vita e delle opere del Marc. Luigi Marini. Discorso del Cav. Camillo Ravioli*, Tipografia di Tito Ajani, Roma, 1858.

[Miscellanea, 1878] Regia Deputazione sovra gli Studi di Storia Patria per le Antiche Provincie e la Lombardia, in *Miscellanea di storia italiana*, vol. 17, Torino, Fratelli Bocca librai di S. M., 1878.

[Romagnani, 1994] G.P. Romagnani. L'istruzione universitaria in Piemonte dal 1799 al 1814. In [Aquila imperiale, 1994, vol. II, pages 536-569].

[Ruffini, 1807] P. Ruffini Paolo. *Corso di matematiche ad uso degli aspiranti alla Scuola d'artiglieria e genio di Modena Tomo terzo contenente l'Algebra elementare del dottor Paolo Ruffini*, presso la Società tipografica, Modena, 1807.

[Steele, 1994] B.D. Steele. Muskets and Pendulums: Benjamin Robins, Leonhard Euler, and the Ballistics Revolution. *Technology and Culture*, 35/2: 348-382, 1994.

[Taton, 1986] R. Taton, editor. *Enseignement et diffusion des sciences en France, au XVIIIe siècle*, Hermann, Paris, 1986.

[Trombetta, 2011] V. Trombetta. *L'editoria a Napoli del decennio francese. Produzione libraria e stampa periodica tra Stato e imprenditoria privata (1806-1815)*, Franco Angeli, Milano, 2011.

[Velamazán and Ausejo, 2020] M. Á. Velamazán and E. Ausejo. La formation mathématique des ingénieurs militaires en Espagne au XIXe siècle. In [Blanco and Bruneau, 2020, pages 9-26].

[Verin, 1993] H. Verin. *La gloire des ingénieurs. L'intelligence technique du XVIe au XVIIIe siècle*, Albin Michel, Paris, 1993.

Pasqual Calbó Caldés, a Minorcan Artist-Scientist, and His Mathematical Course (c. 1800)

Antoni Roca Rosell

Universitat Politècnica de Catalunya – Barcelona Tech / Institut d'Estudis Catalans, Spain

antoni.roca-rosell@upc.edu

Abstract

Pasqual Calbó Caldés (1752-1817) combined his work and talent as painter and draughtsman with the teaching of a course of mathematics. Calbó lived in Minorca and spent nine years in Venice, Rome and Vienna for his training as artist, during which time he almost certainly took advantage of this stay to learn mathematics, i.e. pure and mixed mathematics, probably for the purpose of developing his ability for perspective or architecture. Calbó prepared a course for teaching "young Minorcan artisans", a manuscript of which consisting of around 500 fols., written in Minorcan Catalan is preserved. This manuscript contains pure mathematics (decimal fractions, algebra, geometry, trigonometry and logarithms), experimental physics, sundials, perspective, architecture, and boat-building. The text provides us with a striking example of private technical education. As is well known, the schools of technical education were in process of consolidation during the second half of the XVIII century. Mathematics (including mixed mathematics) were regarded as the basis of technical education and played an important role in these early schools. Engineers, architects and artisans acquired their mathematical training through private courses and handbooks, usually drawn up by private teachers or persons in charge of exams, with the intention of joining the army or obtaining technical qualifications. Calbó's course is currently being

This paper is included in the research project of the Spanish Ministry of Science and Innovation PID2020-113702RB-I00 studied and it seems that he prepared his text by using textbooks by Tomàs-Vicent Tosca, Giorgio Saverio Poli, and Benet Baïls. These textbooks were used for private study and for the training of architects in Spain, and for the training of doctors and engineers in Venice and Naples. This course is an interesting example of private technical education, generally not available, but but it was a way of training that probably highly circulated throughout Europe around 1800.

1 Pasqual Calbó Caldés, an artist-scientist

Pasqual Calbó Caldés (1752-1817) was an outstanding painter[1] born in Maó[2] (Minorca). At the time of his birth and during much of his life, Minorca was under British administration[3] as a result of the Peace of Utrecht (1713), which marked the end of the War of Spanish Succession (1700-1715).[4] The young Pasqual received an artistic education in Maó thanks to the Italian painter Giuseppe Chiesa Barati (1720-1789), who lived in the town where he occupied the post of Vice-consul of the Grand Duchy of Tuscany. Probably on the advice of Chiesa Barati, Pasqual went to Italy in 1770 to further his training as an artist. He spent four years in Venice and four years in Rome, with a grant from Wenzel Anton Fürst von Kaunitz-Rietberg (1711-1794), Chancellor of the Austrian Empire. In 1779, he went to Vienna where he was appointed draughtsman of the new Belvedere Art Gallery, under the auspices of the Empress Maria Theresa. Nevertheless, after some months, in 1780, he decided to return to Minorca.

There is some evidence of Pasqual Calbó's life in Italy and Austria, mainly thanks to Kaunitz's correspondence with Calbó's tutors. One

---

[1]On the life and contributions of Pasqual Calbó, see: [Pons Povedano, 2016]; [Pons Povedano, 2017]; [Andreu Adame and Desel González (eds.), 2017].

[2]Maó is the official Catalan name of the city known in English as Mahon or Port Mahon.

[3]There were three periods of British administration (or occupation): 1713-1756; 1763-1782; 1798-1802; a period of French administration: 1756-1763; and a period of Spanish administration: 1782-1798. In 1802, the island was definitely under this Spanish administration.

[4]The Spanish stage of the war finished in 1715, after the siege of Barcelona (1714) and the surrender of the island of Mallorca (1715).

document preserved in the Public Library of Maó, however, is the reply to a request to the Pope for permission to read forbidden books.[5] This response in May, 1776, is in the name of Pope Pius VI and gives more details on the subjects in which Calbó was interested: politics, philosophy, mathematics, grammar, poetry, rhetoric, and profane and ecclesiastic history. The document authorizes him to read these types of books, but with some exceptions, such as those on astrology and books on history written by Niccolò Machiavelli, in addition to some French Protestant authors (Charles Dumoulin, Pierre Bayle or Isaac de Larrey) as well as the book by Lucretius translated into Italian. These particular references probably concern authors mentioned in Calbó's request, and thus provide us with information about his interests and ambitions that go beyond art.

Calbó's progress in drawing and painting during the Rome years is documented through the correspondence between his tutors and the Chancellor Kaunitz.[6] There are no documents that correspond to his stay in Vienna from 1770 to 1774. Nevertheless, it is reasonable to assume that he was in contact with the Venetian Academy of Fine Arts. We have identified the course on the teaching architecture proposed by Giovanni Gianfrancesco Costa in 1767,[7] which is very similar to some contents of Calbó's writings. Finally, regarding his training as a technician, a very interesting mention exists in one of the passports he needed for his return trip from Vienna to Minorca in 1780. This passport, issued by the French ambassador in Venice, described Calbó and the purpose of his trip by saying that he was "first geographer engineer" and that he was undertaking the journey for his own instruction.[8] In the other passports, Calbó is referred to as "first draughtsman" at the service of the Empire. The reference to him as an "engineer" is somewhat misleading; it appears to be a strategy to facilitate his passage through France, from

---

[5] Document conserved at the Arxiu Històric de Maó and reproduced in [Andreu Adame and Desel González (eds.), 2017, p. 221] (Catalan version, p. 220).

[6] There is a detailed analysis of this correspondence in: [Sintes y de Olivar, 1987]. See also: [Pons Povedano, 2016], [Pons Povedano, 2017].

[7] [Ceccon, 2012].

[8] "Paschal Calbo, premier Ingenieur Geographe de leurs $M.M.II^{les}$ et $R.^{ls}$ [Magestés Impériales et Royales], voyageant pour son Instruction allant en France", ref. U_396_10, Arxiu Històric de Maò.

where he would probably to take a ship to Minorca.

From all of the foregoing, we may conclude that Calbó was able to acquire a comprehensive education, not only in art but also in science and technology. Later, just before his death in 1817, he replied to Joan Ramis, who was compiling a glossary of prominent Minorcans,[9] that he had devoted some of his time to teaching because of his bad health. He said that he had taught "drawing, practical geometry, and civil architecture".[10] What he described as an occasional activity was in fact a manuscript of some 500 pages that is preserved at the *Museu de Menorca* in Maó.

2 Artisans, practical technicians, regular training, and science and technology

Many authors have written about the significant role of artisans and craftsmen in the Scientific Revolution. Artisans contributed a particular knowledge of Nature, generally in terms of practice and technique. Edgar Zilsel regarded craftsmen as artist–engineers who played an essential role in the origin of modern science; Paolo Rossi analysed the role of "machines" in the new vision of nature, while other authors such as Peter Dear, Lissa Roberts and Liliane Hilaire-Perez[11] have delved more deeply into this subject. Modern science and technology were not simply the result of theoretical arguments, and it is necessary to highlight the importance of practice (or practices) in this matter.

With regard to technology, it was very important to have the suitable background for developing abilities and projects. Until the generalisation of regular training in schools, engineers and architects received a practical education directly from other engineers and architects. This is what Melvin Kranzberg and other authors have referred to as "workshop culture".[12] It was necessary for such practical training to be comple-

---

[9][Ramis i Ramis, 1817]

[10]"Al principi vaig exercitar la pintura sols a temporades per causa de la mia poca salut, i en seguida he ensenyat disseny, geometria pràctica i arquitectura civil". Reproduced in [Calbó i Caldés, 2020, pp. 607–608].

[11][Zilsel, 2003]; [Rossi, 1962]; [Dear, 1995], [Dear, 2001]; [Roberts et al, ed., 2007]; [Long, 2011]; [Hilaire-Pérez, 2016].

[12][Kranzberg (ed.), 1986].

mented by theoretical resources. In the case of engineering and architecture, these resources were denominated as "mathematics", including pure and mixed mathematics.[13] A new scientific literature emerged in the form of textbooks under the general title of "mathematical courses", some of which were mainly printed in the XVII and XVIII centuries. The mathematical content included pure and mixed mathematics; pure mathematics consisting of arithmetic, geometry and algebra, as well as calculus in the late XVIII century, while mixed mathematics, sometimes called physic mathematics, consisted of statics, laws of the motion, hydrostatics, hydraulics, artillery, military and civil architecture, optics, astronomy, cosmography, geography, cartography, gnomonics, and (in the late XVIII century) experimental physics, among other subjects. Mixed mathematics therefore consisted of a set of very different disciplines (in our view), all of them subjects in which the pure mathematics were used. In the late XVIII century, experimental physics, as part of natural philosophy, was also included in mixed mathematics.[14]

Some of this material appeared in print, of which Pierre Hérigone's course is an early example.[15] Some printed courses were really appreciated, such as those by Jacques Ozanam,[16], Claude François Milliet Dechales[17], Bernard Forest de Bélidor and Christian Wolff. There also were several Spanish mathematical courses.[18] In the XVIII century, the most influential were those drawn up by Tomàs–Vicent Tosca, of which nine volumes were printed between 1707 and 1715,[19] with at least three reprints during the century, and that by Benet Baïls, with 11 volumes published between 1779 and 1784.[20] In 1756, a Chair of Mathematics was founded in Barcelona, the initiative for which came from the Society of Jesus and one its members, Tomàs Cerdà, who was then a teacher at the noble college of Barcelona. This chair, with a grant from the Spanish Crown and supported by the Municipality, was not exclusively

---

[13][Massa-Esteve et al., 2011]; [Dear, 2011].

[14][Heilbron, 1980], [Heilbron, 2011].

[15][Massa-Esteve, 2008].

[16][Ozanam, 1695].

[17][Dechales, 1674].

[18][Garma Pons, 2002]; [Roca-Rosell, 2019].

[19][Tosca, 1707-1715]. On this author: [Navarro Brotons, 2003], [Navarro Brotons, 2008].

[20][Bails, 1779–1804]. On his contributions: [Martínez Verdú, 2017].

for nobles but also open to all types of citizens. Tomàs Cerdà prepared a mathematical course from which he published sections on arithmetic and geometry. The lessons on artillery were also published, although the rest of course remained in manuscript form and is preserved at the Spanish Royal Academy of History in Madrid, after the expulsion of the Jesuits in 1767.[21] A manuscript course prepared by Pedro de Lucuce for the Military Academy of Barcelona also existed, which served as the main text for military training during the middle of the XVIII century.[22]

Regular technical training in schools was the result of a relatively long process, during which different options were available. Here we provide a model consisting of four steps:[23]

1) Training is a result of the personal contact between master and disciples. This was characteristic of the guilds and also military or civil engineers and architects. Moreover, this method of learning still currently exists for certain technologies; for example, in computing sciences and technology.

2) Training is conducted in private courses or through personal study in handbooks. These courses used to be organised in parallel with the colleges or universities in which mathematical education was largely very limited. In general, the objective of these private courses was to prepare candidates wishing to join the Army or other corps for exams, as well as to complement the training of engineers and architects.

3) Some of these private courses were included in colleges or universities, although they did not form a regular part.

4) The courses were normally taught in institutions established to provide this specific technical education. These new centres were called schools (initially not linked with the colleges or universities).

These four steps constitute only an outline of a series of rapidly changing situations, but we believe that they facilitate understanding of

---

[21] As part of History of Science projects, two treatises by Cerdà have recently been published: [Cerdà, 1999]; [Cerdà, 2015]. See also [Berenguer, 2016].

[22] [Massa-Esteve et al., 2011].

[23] [Roca-Rosell, 2019]; [Roca-Rosell, 2021].

the process from workshop to the technical culture of the schools. This process culminated in the middle of the XIX century, when most of the engineers and architects were trained in regular schools. Nevertheless, some early schools existed during the XVI and XVII centuries. Brice Cossart has recently shown that a school for artillery was established in Seville in the late XVI century,[24] and it is well known that Simon Stevin promoted a school of engineering at the University of Leiden in the early XVII century.[25] In Portugal, the Jesuits introduced some engineering training in its *Aula da Esfera*. Luis Serrão Pimentel was educated there and promoted the "true" Portuguese engineering, which in this case was linked to the Army.[26] In France, the Navy and the Army set up academies for the training of officers.[27] During the XVIII century, some schools were founded in Europe, the French schools being of particular importance. In 1747, under the supervision of Jean–Rodolphe Perronet, a *Bureau de dessinateurs* belonging to the *Corps of Ponts et Chaussées* constituted the origin of a civil engineering school that became fully established in 1775. The School of Military Engineering in Mézières was founded in 1748, which deserves to be recognized as the first regular engineering school in France. Thus, in the second half of the XVIII century, steps 3 and 4 existed simultaneously (with steps 1 and 2).

In Spain, the first civil engineering school was established in Almadén in 1777 as an academy of mines, with the academies of Freiberg and Prague as models.[28] On the basis of the Paris model, a school of *Puentes y Canales* was founded in 1802 on the initiative of Agustín de Betancourt.[29]. Nevertheless, some earlier military schools already existed, such as the Academy of Mathematics in Bruxelles, promoted in 1675 by Sebastián Fernández de Medrano; the school for marine officers in Cadiz (1715);[30] the Barcelona Academy of Mathematics (1720)[31] and

---

[24][Cossart, 2021].

[25][Gouzévitch; Grelon; Karvar (coord.), 2004].

[26][Conde, Massa-Esteve, 2018].

[27]See the contributions in [Blanco and Bruneau (eds.), 2020].

[28][Silva Suárez, 2005]; [Gouzévitch & Gouzévitch, 2011].

[29][Gouzévitch, 2018].

[30][Silva Suárez, 2005].

[31][Capel Saéz et al., 1988]; [Massa-Esteve et al., 2011]; [Roca-Rosell, Massa-Esteve, 2020].

the school of artillery in Segovia (1764).[32] Thus, in Spain step 4 was accomplished in the last third of the XVIII century, while steps 2 and 3 were already running.

3 Calbó's mathematical course

Calbó's manuscript consists of a mathematical course containing the usual matters addressed in such courses.[33] It should be pointed out that these mathematical courses were written in non-academic languages (in Herigone's course, Latin and French, printed in two separate columns) in order to make them accessible to non-academic readers. Jacques Ozanam believed that his course was superior to others written in "foreign" languages such as Latin.[34] Calbó's course is written in Minorcan Catalan, which in fact was unique in Catalan culture at that time. It is necessary to remember that, after the end of the War of Spanish Succession, the Bourbon King Philip V dissolved the statutes of the old reigns in Spain.[35] In 1716, a Royal Decree established the institutions of Castile, suppressing those in Catalonia and prohibiting the official and public use of the Catalan language. However, in Minorca, which had become a British possession, the Catalan language continued to be the main means of communication.[36]

Calbó's manuscript consists of 12 treatises ("tractats") on pure mathematics: decimal fractions, proportions, algebra, logarithms, geometry, solids, plane and spherical trigonometry. After these, there are 4 treatises on experimental physics (mechanics, optics, electrostatics, astronomy, gases, meteorology, optics, acoustics); a treatise on sundials; a treatise on perspective; a treatise on civil architecture and another on military architecture, and finally a treatise on shipbuilding.[37]

---

[32] [Silva Suárez, 2005].

[33] [Roca-Rosell, 2016a], [Roca-Rosell, 2016b], [Roca-Rosell, 2017], [Roca-Rosell, 2020].

[34] [Ozanam, 1695].

[35] See, for example, [Vilar, 1977].

[36] [Murillo, 1986]; [Casasnovas, 2005]. See also: [Gregory, 1990].

[37] We have divided the manuscript into three parts (pure mathematics, experimental physics and the other treatises) for the printed edition in three volumes. Two volumes are printed: [Calbó i Caldés, 2020]; [Calbó i Caldés, 2022].

Some sources have been identified in Calbó's course. Despite the fact that there is no mention of authorship, Vicente Meavilla has shown that some texts in pure mathematics are translations from Tosca, Cerdà, and Baïls.[38] In fact, in his reply to Joan Ramis, Calbó said that his treatises had been translated from "different authors", but without mentioning them by name. Although there are some translations of texts by those authors, several aspects of Calbó's organization of those matters are worthy of remark. For example, in his introduction to calculations, he began with decimal fractions, but without any reference to natural numbers or fractions in general. It is known from Simon Stevin that decimal fractions were identified by artisans as being the most convenient form of numbers for calculation, measurement or design ([Goldstein, 2017]). Geometry and practical geometry constitute more or less half of the section devoted to pure mathematics. One of Calbó's few comments in this regard reads as follows:

> Many figures or operations are useful for various arts and crafts; in fact, there is almost no trade in which the *menestral*,[39] meditating on his works, does not undertake operations that geometry facilitates and orders with rule and proportion.[40]

The treatise on Geometry, he wrote, is dedicated to the "young Minorca's menestrals" from a Minorcan interested in the progress of these young men.[41] Calbó's remark on the need for craftsmen to acquire mathematical knowledge in order to "meditate" their works, providing them with "rule and proportion", is certainly interesting.

The section devoted to experimental physics occupies a third of the manuscript and includes statics, mechanics, cosmology, electricity, heat, gases, meteorology, optics, catoptrics and sound. It is here that we find another way of proceeding. In order to develop these subjects, Calbó chose to follow the handbook on experimental physics written by the Italian naturalist and physicist Giuseppe Saverio Poli,[42] who was

---

[38][Meavilla, 2020].
[39]A particular Catalan word for craftsman.
[40]Translation by the author of this paper.
[41][Calbó i Caldés, 2020, p. 227].
[42]First edition: [Poli, 1781].

teacher in Naples. He was a scientist of some international standing and was a fellow of the Royal Society of London.[43] Most of Calbó's treatises were probably taken from the 1804 edition of this textbook by Poli.[44] It should be pointed out that Poli published more than 20 editions between 1781 and 1824, usually updated with the latest contributions by the author. We settled on the 1804 edition by comparing Calbó's version with the different editions of Poli's book.

While no reference is made to Poli at the beginning of the treatise on experimental physics, it is possible to compare Calbó's text with the corresponding texts by Poli paragraph by paragraph. This shows us that Calbó summarized the developments, bypassing most of Poli's references and avoiding the development of the theoretical arguments. In his text, Calbó provides the main definitions and carefully describes the main experiments.

This style is somewhat surprising. It may seem strange to a reader in the XXI century that while following in Poli's footsteps, Calbó presents Newton's laws of motion, but makes no mention of his name. He only refers directly once to Newton when speaking about his study of the optical properties of diamonds (f. 226rbis).

Two other sources are mentioned in the treatises on experimental physics. One refers to a treatise on statics and hydrostatics in accordance with the work by Tosca, an author who Calbó also mentions when following some treatises on optics. In this section, he also refers to Milliet Dechales, while in the case of the translations from Tosca, Calbó does not follow these texts but only takes from them the main definitions and contents. He writes the treatise in his own style, choosing to simplify the theoretical arguments and pointing out the most relevant experiments. In his texts, Tosca, who was a priest, included several references to divine intervention, which Calbó omitted entirely. All the foregoing gives us a good idea of the way in which the theoretical and technological knowledge of his time was acquired.

The third section, currently being prepared for publication, is devoted to different types of construction: sundials, perspective, architec-

---

[43] On Poli, see [Castellani, 1975]; [Borelli, 2015], and the papers of the session devoted to him in: [Società Italiana degli Storici della Fisica e dell'astronomia, 2017].

[44] [Jordi Taltavull et al., 2022].

ture and shipbuilding. It could be said that this section "applies" the other treatises to various kinds of constructions. Many mathematical courses contain a treatise on sundials, a technique known from Classical Antiquity, but renewed during the Renaissance, mainly by Christopher Clavius. Calbó's source is the corresponding treatise by Tosca, Although, as Guerola Olivares [in press] states, Calbó's text is not a simple translation; it reveals a deep understanding of the elements of gnomonics, which he presents in a practical way. It should be remembered that currently in Minorca many sundials can be found in buildings. It is possible that the lessons drawn up by Calbó would have been useful in the design of such sundials.

This third section also includes a detailed treatise on perspective applied to architecture and to shipbuilding. While Tosca was the source of the treatise on military architecture, that was not the case for civil architecture or for the treatise on perspective. To date, we do not know the sources of the treatise on shipbuilding.

As previously stated, the lessons in Calbó's course seem to be aimed at helping Minorcan technicians to improve their work. According to Sintes Espasa, after 1802 the Spanish authorities consulted builders about their qualifications,[45] to obtain which they were required to pass an exam set by the Royal Academy of Fine Arts. These exams were held in Mallorca, and it is clear that the contents of Calbó's course were quite suitable for their preparation.

Pasqual Calbó belonged to the Minorcan Enlightenment group of intellectuals whose members were very active, although their activities were severely curtailed after the Spanish domination in 1802. [46] This group should be understood in the context of the Republic of Letters, the informal European group who shared interests in science, technology and culture.[47] Calbó would have been the "scientist' in the Minorcan group, which had few members since it was located on a small island.

Andreu Murillo analysed the state of education in Minorca at the time when the island came once again under Spanish sovereignty (1782–1798; 1802 henceforth).[48] He says:

---

[45][Sintes Espasa, 2017].
[46][Salord, 2010], [Salord, 2017].
[47][Roche, 1988].
[48][Murillo, 1986].

According to a collective petition dated January 15, 1810, 39 artisans were asking an unidentified "Ilm. Sr.' for a school to be established with public funds for imparting drawing and painting classes, [also] on Commercial Arithmetic, Nautics, modern languages (two classes), Mathematics and Experimental Physics, as well as a class for teaching the second part of Latin Grammar and Rhetoric.[49]

Beginning in 1803, Karl Ernest Cook, a German teacher living in Maó, set up an elementary school in Maó that constituted a renewal of the education system on the island, which until then had been in the hands of various convents. Although Cook departed Minorca in 1805,[50] his legacy was taken up by several outstanding Minorcans, such as Mateu Orfila.

It is interesting to reflect on the fact that Cook's school and Calbó's classes were probably running simultaneously and provided a valuable alternative to the limited education available on the island at that time. In that sense, the proposal made by these 39 artisans in 1810 seems also to be an appeal for the recognition of Calbó's endeavours and the recovery of the work carried out by Cook. Unfortunately, the state of education in Minorca in general did not appreciably improve, and according to Motilla Salas ([2005]) it was not until the second half of the century that the situation began to change.

4 Conclusion

Calbó's manuscript constitutes a singular testimony to the mathematical training c. 1800. It belongs to the tradition of mathematical courses composed of pure and mixed mathematics. The course has a practical orientation based on the resolution of problems, the practice of experiments or the instructions for building different things (design of sundials,

---

[49]"Segons consta en una petició collectiva de 15 de gener de 1810, 39 menestrals demanaven a no s'expressa quin Ilm. Sr.- que amb fons públics s'establís una escola amb aules de dibuix i pintura, d'Aritmètica comercial, de Nautica, dues de llengües modernes, una de Matemàtiques i Física experimental, una per a l'ensenyament de la segona part de la Gramàtica llatina i Retòrica.", [Murillo, 1986, pp. 78–79].

[50][Murillo, 1986].

architecture, shipbuilding). The course is a very interesting example of step 2 of technical education, in which private lessons helped technicians to improve their training in order to enter to the Army or to obtain some qualifications.

As a member of the Minorcan Enlightened group, through his teaching activity Pasqual Calbó joined the Republic of Letters that was engaged in the renewal of science, technology and technical education in Europe, of which his choice of Giuseppe Saverio Poli's textbook is a prime example.

Given its practical orientation, Calbó wrote the text in the vernacular language of Minorca, a variety of Catalan, which constitutes a singular occurrence in Catalan and Spanish history.

As in many similar cases, Calbó's course never appeared in print. However, we are at present engaged in a critical edition as a way of preserving and interpreting of a scientific heritage. It seems that Calbó's approach to teaching was discontinued after his death in 1817, although some technical education was pursued on the island a few years later. Calbó may only have been a minor figure as scientist, but he contributed to the dissemination and acquisition of new science and technology.

References

[Andreu Adame and Desel González (eds.), 2017] C. Andreu Adame, and C. Desel González, editors. *Pasqual Calbó i Caldés 1752-1817*. Museu de Menorca, Maó, 2017.

[Bails, 1779–1804] B. Baïls. Elementos de matematica, 11 vol. Imprenta por D. Joachín Ibarra, Madrid 1779-1804.

[Berenguer, 2016] J. Berenguer Clarià. *La Recepció del càlcul diferencial a l'Espanya del segle XVIII. Tomàs Cerdà: introductor de la teoria de fluxions*. Universitat Autònoma de Barcelona. Barcelona, 2016. https://www.tdx.cat/handle/10803/367217#page=1

[Blanco and Bruneau (eds.), 2020] M. Blanco and O. Bruneau (eds). Les mathématiques dans les écoles militaires (XVIIIe-XIXe siècles). *Philosophia Scientiae*, 24/1, 2020. https://journals.openedition.org/philosophiascientiae/2122

[Borelli, 2015] A. Borelli. Poli, Giuseppe Saverio. In *Dizionario Biografico degli Italiani*. p. 84, 2015. http://www.treccani.it/enciclopedia/giuseppe-saverio-poli_%28Dizionario-Biografico%29/

[Brown, 1991] G. I. Brown. The Evolution of the Term 'Mixed Mathematics'. *Journal of the History of Ideas*, 52/1: 81–102, 1991.

[Calbó i Caldés, 2020] P. Calbó i Caldés. *Obra científica, I: Tractats de matemàtiques pures*. A. Roca Rosell, J. Salord, i J.-Ll. Torres (coordinators), M.Toldrà i Sabaté, ed. Institut Menorquí d'Estudis; Departament de Cultura, Educació i Esports del Consell Insular de Menorca; Institut d'Estudis Catalans; Universitat de les Illes Balears; Institut d'Estudis Baleàrics; Institut d'Indústries Culturals de les Illes Balears, Maó, Barcelona, Palma, 2020.

[Calbó i Caldés, 2022] P. Calbó i Caldés. *Obra científica, II: Tractats de física experimental i matemàtica*. A. Roca Rosell, M. Jordi i J.-Ll. Torres (coordinators), M.Toldrà i Sabaté, ed. Institut Menorquí d'Estudis; Departament de Cultura, Educació i Esports del Consell Insular de Menorca; Institut d'Estudis Catalans; Universitat de les Illes Balears; Institut d'Estudis Baleàrics; Institut d'Indústries Culturals de les Illes Balears, 2022.

[Calbó i Caldés, in press] P. Calbó i Caldés. *Obra científica, III: Tractats de construccions*. In press.

[Capel Saéz et al., 1988] H. Capel Sáez; J-E. Sánchez, O. Moncada. *De Palas a Minerva: la formación científica y la estructura institucional de los ingenieros militares en el siglo XVIII*. Ediciones del Serbal, CSIC, Barcelona, Madrid, 1988.

[Casasnovas, 2005] M.-A. Casasnovas. *Història de Menorca*. Moll, Palma, 2005.

[Castellani, 1975] C. Castellani. Poli, Giuseppe Saverio. In: Ch. C.Gillispie (ed.) *Dictionary of Scientific Biography*, 11: 66-67, Charles Scribner's Sons, New York, 1975.

[Ceccon, 2012] E. A. Ceccon. *I concorsi di architettura all'Accademia di Belle Arti di Venezia*. Università Ca' Foscari di Venezia. Dipartimento di Filosofia e Beni Culturali, Venice, 2012.

[Cerdà, 1999] T. Cerdà. *Tratado de astronomía: curs dictat l'any 1760 a la Reial Càtedra de Matemàtiques del Col·legi de Sant Jaume de Cordelles, inspirat en la Philosophia Britannica de Benjamin Martin*. L. Gassiot (ed.). Reial Acadèmia de Ciències i Arts de Barcelona, Barcelona, 1999.

[Cerdà, 2015] T. Cerdà. *Tratado de fluxiones: 1757-1759*. J. Berenguer (ed.). Reial Acadèmia de Ciències i Arts de Barcelona, Barcelona, 2015.

[Conde, Massa-Esteve, 2018] A. F. Conde; M. R. Massa-Esteve. Teaching Engineers in the Seventeenth Century: European Influences in Portugal. *Engineering Studies*,10, 2-3: 115–132, 2018.

[Cossart, 2021] B. Cossart. *Les Artilleurs et la Monarchie hispanique (1560-1610). Guerre, savoirs techniques*. État. Classiques Garnier, Paris, 2021.

[Dear, 1995] P. Dear. *Discipline and Experience: The Mathematical Way in the Scientific Revolution.* The University of Chicago Press, Chicago; London, 1995.

[Dear, 2001] P. Dear. *Revolutionizing the Sciences: European Knowledge and its Ambitions, 1500-1700.* Palgrave Macmillan. New York, 2001.

[Dear, 2011] P. Dear. Mixed Mathematics. In Harrison et al. (eds.), *Wrestling with Nature*, pages 149-172, University of Chicago Press, Chicago, London, 2011.

[Dechales, 1674] C-F. M. Dechales. *Cursus seu Mundus Mathematicus, 3 vol.* Ex officina Anissoniana, Lyon, 1674.

[Garma Pons, 2002] S. Garma Pons. La enseñanza de las matemáticas. In J. L Peset Reig (ed.). *Historia de la Ciencia y de la Técnica en la Corona de Castilla, IV: El siglo XVIII*, pages 311–346. Junta de Castilla y León. Consejería de Educación y Cultura, Valladolid, 2002.

[Goldstein, 2017] C. Goldstein. Les fractions décimales: un art d'ingénieur? In R. Carvais, A-F. Garçon, A. Grelon (eds). *Penser la technique autrement XVI-XXI siècle. En homage à l'œuvre d'Hélène Vérin*, pages 185–203. Classiques Garnier, Paris, 2017. Author version: `https://hal.archives-ouvertes.fr/hal-00734932v2`

[Gouzévitch & Gouzévitch, 2011] I. Gouzévitch and D. Gouzévitch. Augustin Betancourt and Mining Technologies. From Almadén to St Petersburg (1783–1824). *History of Technology.* 30: 13–31, 2011.

[Gouzévitch, 2018] I. Gouzévitch. *Planète Bétancourt.* Université Paris-Diderot, 2018.

[Gouzévitch; Grelon; Karvar (coord.), 2004] I. Gouzévitch; A. Grelon; and A. Karvar, coordinators. *La formation des ingénieurs en perspective: modèles de référence et réseaux de médiation XVII–XX siècles.* Presses universitaires de Rennes, Rennes, 2004.

[Gregory, 1990] D. Gregory. *Minorca, the Illusory Prize: A History of the British Occupations of Minorca Between 1708 and 1802.* Fairleigh Dickinson University Press. 1990.

[Guerola Olivares, in press] J. Guerola Olivares. El tractat disetè de l'Obres Didàctiques de Pasqual Calbó i Caldés: Els rellotges de Sol. In Calbó i Caldés, *Obra científica 3*, in press.

[Heilbron, 1980] J. L. Heilbron. Experimental natural philosophy. In G. S. Rousseau; R. Porter, editors. *The Ferment of Knowledge. Studies in the Historiography of Eighteenth Century Science*, pages 357–387. Cambridge University Press, Cambridge, 1980.

[Heilbron, 2011] J. L. Heilbron. Natural Philosophy. In Harrison et al., ed.,

Wresting with Nature, pages 173–199, University of Chicago Press, Chicago, London, 2011.

[Hilaire-Pérez, 2016] L. Hilaire-Pérez. L'artisan, les sciences et les techniques (XVIe-XVIIIe siècle. In L. Hilaire-Pérez; F. Simon; M. Thébaud-Sorger, editors, *L'Europe des sciences et des techniques. Un dialogue des savoirs, XVe-XVIIIe siècle*, pages 103-110. Presses Univesitaires de Rennes, Rennes, 2016.

[Jordi Taltavull et al., 2022] M. Jordi Taltavull, A. Roca-Rosell, S. Vallmitjana Rico. La física experimental i matemàtica de l'Obra científica de Pasqual Calbó. In Calbó i Caldés, *Obra científica II*, pages 19–165, 2022.

[Kranzberg (ed.), 1986] M. Kranzberg, editor. *Technological education technological style*. San Francisco Press, San Francisco, 1986.

[Long, 2011] P. O. Long. *Artisan/Practicioners and the Rise of the New Sciences, 1400-1600*. Oregon State University Press, Corvallis, 2011.

[Martínez Verdú, 2017] D. Martínez Verdú. La Concepción didáctico-cognitiva de la enseñanza de las matemáticas en Benito Bails (1731-1797). In P. Grapí Vilumara; M. R. Massa Esteve, editors. *Actes de la XV Jornada sobre la Història de la Ciència i l'Ensenyament*, pages 115-128. SCHCT-IEC, Barcelona, 2017.

[Massa-Esteve, 2008] M. R. Massa-Esteve. Symbolic language in early modern mathematics: the algebra of Pierre Hèrigone (1580-1643). *Historia mathematica*. 35–4: 285-301, 2008.

[Massa-Esteve et al., 2011] M.R. Massa-Esteve; A. Roca-Rosell; and C.Puig-Pla. "Mixed" Mathematics in engineering education in Spain: Pedro Lucuce's course at the Barcelona Royal Military Academy of Mathematics in the eighteenth century. *Engineering Studies*, 3/3: 233–253, 2011.

[Meavilla, 2020] V. Meavilla. Els tractats de matemàtiques pures en l'Obra científica de Pasqual Calbó: estudi introductori. In: Calbó i Caldés, *Obra científica I*, pages 59–113, 2020.

[Motilla Salas, 2005] X. Motilla Salas. Els orígens de l'ensenyament secundari públic a Menorca: de l'Escola de Nàutica a l'Institut Lliure de Segon Ensenyament (1855-1869). *Educació i Cultura*, 18: 23–38, 2005.

[Murillo, 1986] A. Murillo. La crisi educativa arran de la devolució definitiva de Menorca a la Corona d'Espanya (1802). *Educació i Cultura: revista mallorquina de Pedagogia*, 5: 71–82, 1986. https://raco.cat/index.php/EducacioCultura/article/view/70209

[Navarro Brotons, 2003] V. Navarro Brotons. Tradition and Scientific Change in Early Modern Spain: The Role of the Jesuits. In M. Feingold, editor. *Jesuit Science and the Republic of Letters*, pages 331-387. The MIT Press, Cambridge, Massachusetts, London, 2003.

[Navarro Brotons, 2008] V. Navarro Brotons. Tomás Vicente Tosca. In *Real Academia de la Historia. Diccionario Biográfico Español,* 2008. http://dbe.rah.es/biografias/9029/tomas-vicente-tosca

[Ozanam, 1695] J. Ozanam. *Cours de Mathematique qui comprend toutes les parties de cette Science les plus utiles et les plus nécessaires à un Homme de Guerre et à tous ceux qui veulent se perfectionner dans les Mathématiques,* 5 volumes. Jombert, Paris, 1695.

[Poli, 1781] G. S. Poli. *Elementi di fisica sperimentale, composti per uso della regia università dal tenente [...].* Presso Giuseppe Campo, Napoli. [There are some editions in Naples and Venice until 1837, with 2, 4 or 5 volumes. See: *Elementi [...],* 5 vols., Andrea Santini, Venice, 1804; *Elementi [...],* 5 vols, Giustino Pasquali q. Mario, Venice:1804].

[Pons Povedano, 2016] M. A. Pons Povedano. *El Gran Tour del pintor Calbó.* Institut Menorquí d'Estudis; Consell Insular de Menorca: Departament de Cultura i Educació, Maó, 2016.

[Pons Povedano, 2017] M. A. Pons Povedano. Retrat biogràfic del pintor Calbó. In Andreu Adame; Desel Gonzaléz (eds.), *Pasqual Calbó ,* pages 30–37, 2017.

[Ramis i Ramis, 1817] J. Ramis i Ramis. *Varones ilustres de Menorca y noticia de los apellidos que mas se han distinguido en ella.* En la imprenta de Serra, Mahon, 1817.

[Roberts et al, ed., 2007] L. Roberts, S. Schaffer, P. Dear, editors. *The mindful hand. Inquiry and invention from the late Renaissance to early industrialisation.* Koninklijke Nederlandse Akademie van Wetenschappen, Amsterdam, 2007.

[Roca-Rosell, 2016a] A. Roca-Rosell. *Un Curs Matemàtic a la Menorca de la Illustració. En la commemoració de Pasqual Calbó i Caldés (1752-1817).* Institut d'Estudis Catalans, Barcelona, 2016.

[Roca-Rosell, 2016b] A. Roca-Rosell. La història de la ciència i de la tècnica en clau local. Algunes notes sobre el "Curs matemàtic" de Pasqual Calbó (1752-1817), *Randa.* 77: 43–56, 2016.

[Roca-Rosell, 2017] A. Roca-Rosell. Una obra 'matemàtica' singular. Pasqual Calbó, un artista-científic. In Andreu Adame; Desel González (eds.), *Pasqual Calbó i Caldés 1752-1817,* pages 48-59. Museu de Menorca, Maó, 2017.

[Roca-Rosell, 2019] A. Roca-Rosell. Ingénierie et société en Espagne, XVIII et XIX siècles: influences et relations avec la France, modèles et transferts. Quatre stades, plusieurs rythmes. Projet 2018-2019, *Etudes et Documents,* 18, [3-20], 2019. https://www.cmh.ens.fr/Ingenierie-et-societe-en-Espagne

[Roca-Rosell, 2020] A. Roca-Rosell. Pasqual Calbó i el seu curs matemàtic. In Calbó i Caldés, *Obra científica I*, pages 21-58., 2020.

[Roca-Rosell, 2021] A. Roca-Rosell. Els orígens de l'ensenyament tècnic en escoles. Un nou model? In C. Ferragud and M.R. Massa-Esteve, editors, *Actes de la XVII Jornada sobre la Història de la Ciència i l'Ensenyament "Antoni Quintana Marí"*, pages 29–36, Societat Catalana d'Història de la Ciència i de la Tècnica, Barcelona, 2021. https://publicacions.iec.cat/repository/pdf/00000305/00000063.pdf

[Roca-Rosell, Massa-Esteve, 2020] A. Roca-Rosell; M. R Massa-Esteve. L'Acadèmia Militar de Matemàtiques de Barcelona. L'informe de 1730 de Jorge Próspero de Verboom. *Quaderns d'història de l'enginyeria*, XVIII:159-208, 2020. https://upcommons.upc.edu/handle/2117/336655

[Roche, 1988] D. Roche. *Les Républicains des lettres: gens de culture et lumières au XVIIIe siècle*. Fayard, Paris, 1988.

[Rossi, 1962] P. Rossi. *I filosofi e le macchine: 1400-1700*. Feltrinelli Editore, Milano, 1962.

[Salord, 2010] J. Salord. *La Illustració a Menorca*. Documenta Balear, Palma, 2010.

[Salord, 2017] J. Salord. Pasqual Calbó dins la Menorca Illustrada. In Andreu Adame; Desel González (eds.), *Pasqual Calbó*, pages 22–29, 2017.

[Silva Suárez, 2005] M. Silva Suárez. Institucionalización de la ingeniería y profesiones técnicas conexas: misión y formación corporativa. In M. Silva Suárez, editor. *Técnica e Ingeniería en España. II El Siglo de las Luces. De la ingeniería a la nueva navegación*, pages 165-262. Real Academia de Ingeniería, Institución "Fernando El Católico", Prensas Universitarias de Zaragoza, Zaragoza, 2005.

[Sintes Espasa, 2017] G. Sintes Espasa. *L'Expansió urbanística del Maó illustrat arran de la desamortització de 1798-1808*. Institut Menorquí d'Estudis, Menorca, 2017.

[Sintes y de Olivar, 1987] M. Sintes y de Olivar. *Pascual Calbo Caldés: un pintor menorquín en la Europa ilustrada*. Caja de Baleares "Sa Nostra", Palma, 1987.

[Società Italiana degli Storici della Fisica e dell'astronomia, 2017] Società Italiana degli Storici della Fisica e dell'astronomia. *Atti del XXXVI Convegno annuale / Proceedings of the 36th Annual conference (Napoli 2016)*, Salvatore Esposito (ed.), Pavia University Press, Pavia, 2017. http://archivio.paviauniversitypress.it/oa/9788869520709.pdf

[Tosca, 1707-1715] T. V. Tosca. *Compendio Mathematico: en que se contienen todas las materias mas principales de las ciencias que tratan de la cantidad.*

9 vols., por Antonio Borcazar; Vicente Cabrera, València, 1707-1715.

[Vilar, 1977] P. Vilar. *Spain: a brief history*. Pergamon Press, Oxford, 1977.

[Zilsel, 2003] E. Zilsel. *The Social origins of modern science*. Springer Science+Business Media Dordrecht, 2003.

www.ingramcontent.com/pod-product-compliance
Lightning Source LLC
Chambersburg PA
CBHW071629200326
41519CB00012BA/2221